Low-Rate Wireless Personal Area Networks

...Enabling Wireless Sensors with IEEE 802.15.4™

José A. Gutiérrez

Edgar H. Callaway, Jr.

Raymond L. Barrett, Jr.

Published by
Standards Information Network
IEEE Press

Trademarks and Disclaimers

IEEE believes the information in this publication is accurate as of its publication date; such information is subject to change without notice. IEEE is not responsible for any inadvertent errors.

Library of Congress Cataloging-in-Publication Data

> Gutiérrez, José A., 1967-
>   Low-rate wireless personal area networks: enabling wireless sensors with IEEE 802.15.4
> / José A. Gutiérrez, Ed Callaway, Raymond Barrett.
>   p. cm.
>   Includes bibliographical references and index.
>   ISBN 0-7381-3557-7
>   1. Wireless LANs. 2. Sensor networks. I. Callaway, Edgar H. II. Barrett, Raymond,
>   1944-III. Title.
>
>   TK5105.78.G88 2003
>   004.6 8—dc21
>
>   200304711

The Institute of Electrical and Electronics Engineers, Inc.
3 Park Avenue, New York, NY 10016-5997, USA

Copyright © 2004 by The Institute of Electrical and Electronics Engineers, Inc.
All rights reserved. Published November 2003. Printed in the United States of America.

No part of this publication may be reproduced in any form, in an electronic retrieval system, or otherwise, without the prior written permission of the publisher.

Yvette HoSang, Manager, Standards Publishing
Jennifer Longman, Managing Editor
Linda Sibilia, Cover Designer

IEEE Press/Standards Information Network publications are not consensus documents. Information contained in this and other works has been obtained from sources believed to be reliable, and reviewed by credible members of IEEE Technical Societies, Standards Committees, and/or Working Groups, and/or relevant technical organizations. Neither the IEEE nor its authors guarantee the accuracy or completeness of any information published herein, and neither the IEEE nor its authors shall be responsible for any errors, omissions, or damages arising out of the use of this information.

Likewise, while the author and publisher believe that the information and guidance given in this work serve as an enhancement to users, all parties must rely upon their own skill and judgement when making use of it. Neither the author nor the publisher assumes any liability to anyone for any loss or damage caused by any error or omission in the work, whether such error or omission is the result of negligence or any other cause. Any and all such liability is disclaimed.

This work is published with the understanding that the IEEE and its authors are supplying information through this publication, not attempting to render engineering or other professional services. If such services are required, the assistance of an appropriate professional should be sought. The IEEE is not responsible for the statements and opinions advanced in the publication.

Review Policy

The information contained in IEEE Press/Standards Information Network publications is reviewed and evaluated by peer reviewers of relevant IEEE Technical Societies, Standards Committees and/or Working Groups, and/or relevant technical organizations. The authors addressed all of the reviewers' comments to the satisfaction of both the IEEE Standards Information Network and those who served as peer reviewers for this document.

The quality of the presentation of information contained in this publication reflects not only the obvious efforts of the authors, but also the work of these peer reviewers. The IEEE Press acknowledges with appreciation their dedication and contribution of time and effort on behalf of the IEEE.

To order IEEE Press Publications, call 1-800-678-IEEE.

Print: ISBN 0-7381-3557-7   SP1131

See other IEEE standards and standards-related product listings at:
http://standards.ieee.org/

Trademarks

Bluetooth is a registered trademark of Bluetooth SIG, Inc.(www.bluetooth.org/).

DeviceNet is a trademark of ODVA (www.odva.org/).

Fieldbus is a registered trademark of the Fieldbus Foundation (www.fieldbus.org/).

IEEE and 802 are registered trademarks of the Institute of Electrical and Electronics Engineers, Incorporated (www.ieee.org/).

IEEE Standards designations are trademarks of the Institute of Electrical and Electronics Engineers, Incorporated (www.ieee.org/).

INCOM is a trademark of Eaton Corporation (www.eaton.com/).

IrDA is a registered trademark of the Infrared Data Association (www.irda.org/).

ZigBee Alliance is a trademark of Philips Corporation (www.zigbee.org/).

Dedication

To my wife, Natacha, for her lovely support and for the inspiration she brings at every instant of my life. Grácias mi vida!

José A. Gutiérrez

To my parents, Pat and Ed, Sr., for the example they set and the guidance they gave; and to my wife, Jan. Without her support and understanding, this book would not have been possible.

Ed Callaway

To my wife, Julie, and our two wonderful children, Gloria and Raymond, who bring the meaning to my life and works.

Ray Barrett

Acknowledgment

Behind this book is the work of the larger IEEE 802® community, in particular, the outstanding work performed by the technical editorial team of the IEEE 802.15.4™ Task Group, led by the indomitable and indefatigable Dr. Bob Heile.

We would like to express our sincere thanks to a diverse group of persons who provided their support during the standardization process, and with whom we have had many enriching technical discussions: Anthony Allen, Venkat Bahl, Daniel Bailey, Monique Bourgeois (IEEE Std 802.15.4 MAC Technical Editor), Priscilla Chen, Dr. Neiyer Correal, David Cypher, Ralph D'Souza, Dr. Larry Dworsky, Paul Gorday, David Hayes, Dr. Lance Hester, Dr. Ivan Howitt, Dr. Jian Huang, Dr. Yan Huang, Phil Jamieson (IEEE Std 802.15.4 Vice Chair), Pat Kinney (IEEE Std 802.15.4 Chair), Masahiro Maeda, Dr. Fred Martin, Dave McClanahan, Vinay Mitter, Said Moridi (IEEE Std 802.15.4 PHY Technical Editor), Dr. Bob O'Dea, Luis Pereira, Dr. Robert Poor, Larry Puhl, Phil Rudland, Dr. Chip Shanley, Qicai Shi, Bob Stengel, Rene Struik, David Taubenheim, and Hans van Leeuwen. In addition to many of the above, Farron Dacus, Barbara Doutre, Latonia Gordon, Charles Luebke, and Marco Naeve (IEEE Std 802.15.4 Secretary) commented on the many early drafts of this book.

We also would like to thank the anonymous peer reviewers of this book for their evaluation and guidance in helping to enrich the material presented here.

Finally, we would like to thank Jennifer Longman, Angela Ortiz, and the rest of the IEEE editorial team for their diligence and the great work performed.

Authors

José A. Gutiérrez is Principal Engineer at the Innovation Center of Eaton Corporation, leading the technical activities in Wireless Sensor Networks. He received the B.S. degree in electronic engineering from Universidad Simon Bolivar in Caracas, Venezuela, in 1991, and the M.S. in electrical engineering from the University of Wisconsin - Milwaukee in 2001. Currently he is a Ph.D. candidate at the same institution. Mr. Gutiérrez is an active member of the IEEE LAN/MAN Standards Committee and Editor in-Chief of the IEEE 802.15 Working Group, Task Group 4, focused in the development of Low-Rate Wireless Personal Area Networks. His areas of expertise include control systems, wireless communication, networking, information theory, bioelectronics, and standards development. He has two patents filed and is author of several papers in the areas of automatic control, artificial intelligence, and wireless communications.

Edgar H. Callaway, Jr. received the B.S. degree in mathematics and the M.S.E.E. degree from the University of Florida in 1979 and 1983, respectively, the M.B.A. degree from Nova (now Nova-Southeastern) University in 1987, and the Ph.D. degree in computer engineering from Florida Atlantic University in 2002. He joined the Land Mobile Division of Motorola in 1984 as an RF engineer working on trunked radio products, and in 1990 transferred to Motorola's Paging Products Group, where initially he designed paging receivers for the Japanese market. From 1992 to 2000 he was engaged in paging receiver and transceiver system design, and was the lead receiver designer of Motorola's paging platform. In 2000, he joined Motorola Labs, where his interests include the design of low-power wireless networks. He is a Registered Professional Engineer (Florida), and has more than twenty issued U.S. patents and several published papers.

Raymond Louis Barrett, Jr. received the B.S.E.E. degree from Case Institute of Technology (now Case-Western University) in 1966, the M.S. in computer science and M.B.A. degrees from Nova (now Nova-Southeastern) University in 1983 and 1982, respectively, and the Ph.D. in electrical engineering from Florida Atlantic University (FAU) in 1990. He taught as an Assistant Professor for Nova University and FAU, and as an Associate Professor at the University of North Florida from 1982 through 2001. Concurrently with teaching at FAU, he joined the Motorola's Paging Products Group in 1991 as a Senior Member of the Technical Staff whereupon he initially designed various low-power paging receiver circuits. In 1999 he joined Motorola Labs, where his interests included the development of low-power wireless components. He is presently a Principal Member of the Technical Staff with MAXIM Integrated Products working on portable products. He is a Registered Professional Engineer (Florida) and has more than thirty issued U.S. patents and several published papers.

Table of Contents

Introduction xiii

Acronyms and Abbreviations xv

Chapter 1 The IEEE Standard 802.15.4 3
   WLANs, WPANs, and LR-WPANs 4
   ISO/OSI Reference Model 6
   Wireless Sensor Networks 7
      Challenges in the Design of Wireless Sensor Networks 8
      Power Consumption 9
      Range 10
      Availability of Frequency Bands 10
      Network Topology 11
      Self-Organization 11

Chapter 2 Low-Rate Wireless PAN Applications 13
   Industrial and Commercial Control and Monitoring—Wireless Sensors .. 14
   Home Automation and Networking 15
      Consumer Electronics 16
      PC Peripherals 16
      Home Automation 17
      Home Security 17
      Personal Healthcare 18
      Toys and Games 18
   Automotive Sensing 19
   Precision Agriculture 20
   Miscellaneous Applications 21

Chapter 3 IEEE Std 802.15.4 Technical Overview 25
   How Low-Cost and Long Battery Life Are Achieved 25
      Duty cycle 25
      Modulation 26
      Use of Direct Sequence Spread Spectrum 27
      Transmitter power 29
      Receiver sensitivity 29
      Quality of service 29
   Other Notable IEEE Std 802.15.4 Features 29
      Network components 29

       Multiple network topologies. 30
       Channel Access. 31
       Multiple PHYs. 32
       Error control 33
  Four frame types 34
       Beacon frames 34
       Data frames. 35
       Acknowledgment frames 35
       MAC command frame 35

Chapter 4 Physical Layer 37
  Regulatory Considerations 37
       European Regional Case. 38
       North American Regional Case 38
       Worldwide Case 38
  Frequency Bands and Data Rates 39
       Data Rates. 40
  Channel Assignment 40
  PHY Bit Level Communication 41
       2.4 GHz Band Specification. 42
       868/915 MHz Band Specification 44
  Radio Characteristics 47
       Power Outputs 48
       Sensitivity. 48
       Range 48
       Receiver Selectivity 50
       Channel Selectivity and Blocking 51
  PHY Services 51
       PHY Data Services. 52
       PHY Management Services 53
       PHY PAN Information Base Management Primitives. 53
  Enabling and Disabling the PHY 54
  Clear Channel Assessment 54
  Energy Detection 55
  PHY Packet Structure 55

Chapter 5 Medium Access Control Sublayer 59
       Star topology. 60
       Peer-to-peer topology 61
  Superframe Structure 62
  MAC Data Transfer Model 63

      Star Network Data Transfer — Beacon-enabled 64
      Star Network Data Transfer — Non-Beacon-enabled 65
      Peer-to-Peer Data Transfer 66
  MAC Services 67
      Transmission Scenarios 67
      MAC Data Service 70
      MAC Management Service 71
  MAC PAN Information Base Management Primitives 72
  Enabling and Disabling the Receiver 72
  Scanning Radio Channels 73
  Association and Disassociation Control 74
  Guaranteed Time Slot Management 75
  Orphan Device Management 79
  Synchronization Control 79
  Beacon Management 81
  Beaconless Synchronization 83
  Communication Status 83
  MAC Frame Structure 84
      Beacon Frames 86
      Data Frames 86
      Acknowledgment Frame...................................... 87
      MAC Command Frame 87
  MAC Functional Scenarios 88
      Accessing the Channel 88
      Starting and Maintaining PANs 88
      Device Synchronization 89
      GTS Management... 89
  Security Services 90
      Access Control 90
      Data Encryption 90
      Frame Integrity 91
      Sequential Freshness.. 91
      Unsecured Mode.. 91
      Access Control List (ACL) Mode.............................. 92
      Secured Mode.. 92

Chapter 6 Network Functionality 95
  Feature Overview 95
      Starting a Network 95
  Upper Layer Network Formation Policies and Algorithms 98
      Pan Coordinator Selection 98

      Selection of the PAN Identifier 99
      The Use of Beacons 100
  The Star Network 101
  Peer-to-Peer Networks 101
      The Flat "Mesh" Network Topology 102
      The Cluster Network Topology 104
      The Cluster Tree Network Topology 109
  The Network Topology Decision 112

Chapter 7 System Design Considerations 117
      Direct Sequence Spread Spectrum 117
      High channel spacing/modulation bandwidth ratio 117
      Relaxed transmitted error vector magnitude (EVM) requirements ... 118
      Constant-envelope modulation................................ 119
      Relaxed receiver blocking specification...................... 120
  Time and Frequency Reference Tradeoffs 120
  Single-Chip versus Multiple-Chip Implementation 120
  OEM Implementations 121
  Time and Power Management 122
  Antennas 126
  Product Design Flexibility 128

Chapter 8 Real-World Issues 131
  Coexistence 131
      Implementation of Colocated Transceivers 134

Chapter 9 Concluding Remarks 137

References 139

Glossary 143

Index 149

Introduction

Most of the attention of the wireless networking world has been focused on wireless local area networks (WLANs) and the enabling technology driven by IEEE Std 802.11™ [8], [13], [18], [25]. The concept of wireless personal area networks (WPANs) has been around for several years, but it entered the mainstream with the creation of the IEEE 802.15 Working Group. The initial effort was centered on creating the IEEE 802.15.1™ standard based on Bluetooth® wireless technology [14], [20]. With the formation of the IEEE 802.15 Working Group, a large and diverse community of technologists, scientists, and visionaries got together under the IEEE Standards Association (IEEE-SA) umbrella to share their views on short-range wireless networking where little or no infrastructure is required (no network setups and no access points) [1]. The dream was ubiquitous, untethered, short-range communications.

Almost immediately, this community organized into three subgroups with distinct requirements. The first was the original Bluetooth (IEEE Std 802.15.1) community centered on cable replacement for consumer electronic devices (mobile phone centric). The second (IEEE Std 802.15.3™) focused on high-speed WPANs driven by the need for wireless connectivity in multimedia devices such as digital cameras, camcorders, etc. [16].

In December 2000, the IEEE-SA New Standards Committee (NesCom) officially sanctioned the creation of a new project to begin development of a third standard, for low-rate wireless personal area networks (LR-WPANs), to be called IEEE Std 802.15.4™ [15]. The goal of the IEEE 802.15.4 Task Group was to define a standard to provide ultra-low complexity, low-cost, and extremely low-power wireless connectivity among inexpensive fixed, portable, and moving devices. Simplicity was the watchword throughout the standard development process.

IEEE Std 802.15.4 is designed to address applications with relaxed throughput and latency requirements while favoring a low-cost and low-power design. These features enable applications in the fields of industrial, agricultural, vehicular, residential, and medical sensors and actuators, which cannot make use of current wireless technologies or have to use proprietary solutions [2], [7].

This book is designed to complement the IEEE 802.15.4 standard by presenting an overview of the features that characterize it, the applications that motivate it, and the rationale that dictated several of the design options chosen during its creation. The standard is written and structured in a way that makes it very easy to understand and (we believe) easy to implement. The IEEE 802.15.4 technical editing team concentrated especially on creating informative clauses to improve the communication of the more complex concepts found in the standard. This book does not pretend to extend that effort except in discussion of applications, but rather intends to be a companion work. It is directed to people interested in the field of "simple" wireless connectivity. It has a focus in wireless sensors and actuators for the industry in general.

The material in this book is divided into three parts. Part I presents an overview of the low-rate wireless personal area technology and IEEE Std 802.15.4. The material focuses on the motivation for the creation of this standard, including the application scenarios that drove the making of it. Part I is designed not only as a technical introduction to the standard, but also to offer marketing and business professionals enough background information on the technology and the vision behind its conception to help plan marketing and business strategies.

Part II concentrates on the technical features and components of the IEEE 802.15.4 standard. It adds material on network layer functionality that is not covered by the standard. The network layer information applies to the rationale behind the design of several of the components of the standard for the envisioned applications. Part III focuses on implementation and system design considerations. It includes an analysis of system-level real-world issues that will be important for prospective implementers to consider.

The information in IEEE Std 802.15.4 is compact and simplicity is reflected in the text. The standard is organized into seven clauses and seven annexes. Clauses 1 though 5 contain introductory material that maps with Part I of this book. Clauses 6 and 7 in the standard define the physical layer and medium access control (MAC) sublayer, respectively; these clauses map with Part II. Finally, as explained above, Part III does not map to the standard, since it deals with implementation and real-world issues that the standard does not address.

Acronyms and Abbreviations

ACK	acknowledgment
ACL	access control list
AES	advanced encryption standard
BPSK	binary phase shift keying
BSP	beacon synchronization parameter
CAP	contention access period
CBC-MAC	cipher block chaining message authentication code
CCA	clear channel assessment
CCM	counter mode + CBC-MAC
CFP	contention-free period
CRC	cyclic redundancy check
CSMA-CA	carrier sense multiple access with collision avoidance
CTR	counter mode
DSSS	direct sequence spread spectrum
ED	energy detection
ETSI	European Telecommunications Standards Institute
EVM	error-vector magnitude
FCC	Federal Communications Commission
FCS	frame check sequence
FFD	full function device
FHSS	frequency hopping spread spectrum
GSM	global system for mobile
GTS	guaranteed time slot

HVAC	heating, ventilation, and air conditioning
IC	integrated circuit
IrDA	infrared data association
ISM	industrial, scientific, and medical
ITU-T	International Telecommunication Union - Telecommunication Services
LAN	local area network
LC	inductor-capacitor
LLC	logical link control
LQI	link quality indication
LR-WPAN	low-rate wireless personal area network
LSB	least significant bit
MAC	medium access control
MCPS-SAP	MAC common part sublayer service access point
MFR	MAC footer
MHR	MAC header
MIC	message integrity code
MIPS	million instructions per second
MLME	MAC sublayer management entity
MLME-SAP	MAC sublayer management entity service access point
MSB	most significant bit
MPDU	MAC protocol data unit
MSDU	MAC service data unit
MSK	minimum shift keying
OEM	original equipment manufacturer
O-QPSK	offset quadrature phase shift keying
OSI	open systems interconnection

PAN	personal area network
PD-SAP	PHY layer data service access point
PHY	physical layer
PHR	PHY header
PIB	PAN information base
PLME	PHY layer management entity
PLME-SAP	PHY layer management entity service access point
PN	pseudo-noise
POS	personal operating space
PPDU	PHY protocol data unit
PSDU	PHY service data unit
ppm	parts per million
QoS	quality of service
RAM	random access memory
RC	resistor-capacitor
RF	radio frequency
RFD	reduced function device
RFID	radio frequency identification
RSSI	received signal strength indication
RX	receive or receiver
SAP	service access point
SAW	surface acoustic wave
SHR	synchronization header
SNR	signal-to-noise ratio
SoC	system-on-a-chip
SSCS	service specific convergence sublayer

TDMA	time division multiple access
TX	transmit or transmitter
VCO	voltage controlled oscillator
WLAN	wireless local area network
WPAN	wireless personal area network
WSN	wireless sensor network

Part I

Chapter 1   The IEEE Standard 802.15.4

...enabling wireless sensor networks

The explosive growth of embedded control and monitoring in almost any electronic device and the need for connectivity of these applications is causing an integration bottleneck. Manufacturers use different communication interfaces — standard or proprietary — creating the need for yet another controller task to make the arbitration. Conventionally, these communication links are wired. Wires allow power and the reliable transmission of signals from a controller to its peripherals. When the peripherals are not physically contained in the controller, the required wiring brings issues such as cost of installation, safety, and operation convenience to the surface. Wireless technology is the obvious solution to overcome these obstacles, although it comes with its own set of challenges — propagation, interference, security, regulations, and others. The technology to overcome these issues exists, but normally with added complexity causing an increase in the cost of the system.

Certainly some applications can afford the cost of adding a high-end wireless communication system such as cellular phones, IEEE Std 802.11™ Wireless Local Area Networks, and so on. Conversely, many other applications can be enhanced or enabled if a standard low-cost wireless communications solution was available. A standard would drive interoperability among different manufacturers with direct benefit to the end consumer.

A low-rate wireless personal area network (LR-WPAN) is a network designed for low-cost and very low-power short-range wireless communications. This definition is at odds with the current trend in wireless technologies whose focus has been on communications with higher data throughput and enhanced quality of service (QoS).

 The current trend in wireless technologies is to produce protocols with higher data throughput and QoS as their primary performance metrics, with cost and power consumption secondary. The primary performance metrics of an LR-WPAN are cost and power consumption, with data throughput and QoS secondary.

WLANs, WPANs, AND LR-WPANs

Wireless Local Area Networks (LANs) were created as the wireless extension of the IEEE 802® wired LAN (WLAN), which was designed for high-end data networking. Among the system requirements of a WLAN are seamless roaming, message forwarding, longest possible range and capacity for a large population of devices. In contrast, WPANs are designed to function in the Personal Operating Space (POS), extending up to 10 m (30 ft) in all directions and covering the area around a person whether stationary or in motion.

WPANs are used to convey information over relatively short distances among the participant transceivers. Unlike WLANs, connections effected via WPANs involve little or no infrastructure. This allows small, power efficient, inexpensive solutions to be implemented for a wide range of devices.

The IEEE 802.15 Working Group has defined three classes of WPANs that are differentiated by data rate, battery drain, and QoS. The high-data rate WPAN (IEEE Std 802.15.3™) is suitable for multimedia applications that require very high QoS. Medium-rate WPANs (IEEE Std 802.15.1™/Bluetooth®) are designed as cable replacements for consumer electronic devices centered on mobile phones and PDAs with a QoS suitable for voice applications. The last class of WPAN, LR-WPAN (IEEE Std 802.15.4™), is intended to serve applications enabled only by the low power and cost requirements not targeted in the above WPANs. LR-WPAN applications have a relaxed need for data rate and QoS. Figure 1–1 illustrates the operating space of the 802 WLAN and WPAN standards. Notice that IEEE Std 802.15.4 is not designed to overlap with higher end wireless networking standards.

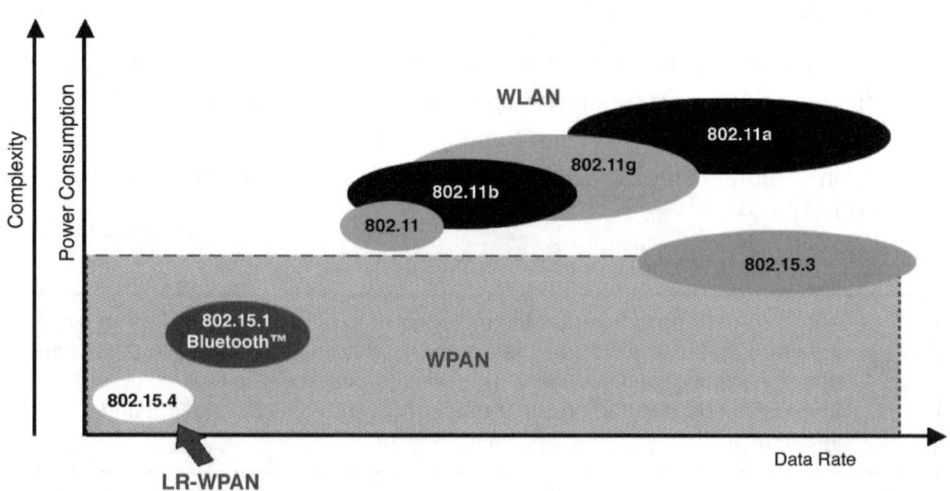

Figure 1–1: Operating space of various WLAN and WPAN standards

Table 1–1 presents a summary of the principal characteristics of an LR-WPAN using IEEE Std 802.15.4, compared with IEEE Std 802.11b™ and a standard WPAN such as IEEE Std 802.15.1.

Table 1–1:   Comparison of LR-WPAN with other wireless technologies

	802.11b WLAN	Bluetooth™ WPAN	Low Rate WPAN
Range	~100 m	~10 - 100 m	10 m
Data Rate	~2-11 Mb/s	1 Mb/s	≤ 0.25 Mb/s
Power Consumption	Medium	Low	Ultra Low
Size	Larger	Smaller	Smallest
Cost/Complexity	Higher	Medium	Very Low

LR-WPAN technology pursues applications where WPAN solutions are still too expensive, extremely low-power operation is needed, and/or the performance of a technology such as Bluetooth is not required.

LR-WPANs complement the range of wireless networking technologies by providing very low-power consumption capabilities at very low implementation cost while enabling applications that were previously impractical or implemented with proprietary technologies.

The intent of IEEE Std 802.15.4 is not to compete with other wireless networking technologies but to complement the range of available wireless technologies in the lower end of the spectrum of data rates, power consumption, and cost.

Although possible for certain applications, IEEE Std 802.15.4 was not designed to overlap its application space with other wireless networking standards.

The authors strongly discourage the use of IEEE Std 802.15.4 for conventional WLAN applications. It may result in a highly challenging endeavor (with very long latencies).

ISO/OSI REFERENCE MODEL

IEEE 802 communication standards define only the bottom two layers of the International Standard Organization (ISO's) Open System Interconnection (OSI) protocol reference model [26]: the physical (PHY) layer and the data link layer [17]. The other layers are not specified in the standard and are normally specified by industrial consortia formed by companies interested in the manufacturing and use of the particular standard.

 The layered reference model allows the encapsulation of different levels of abstraction with a well-defined functionality. Each layer offers services to the layer above through the use of service primitives.

In the case of IEEE Std 802.15.4, the ZigBee™ Alliance is an organization that has led the development of the upper layers, through the definition of application profiles. These profiles make use of a simplified five layer ISO/OSI reference model as shown in Figure 1–2.

	Seven Layer ISO-OSI Model	Simplified Five Layer ISO-OSI Model	IEEE 802 Model
7	Application	User Application	Upper Layers
6	Presentation	Application Profile	
5	Session		
4	Transport		
3	Network	Network	
2	Data Link	Data Link	Logical Link Control (MAC)
			Media Access Control (MAC)
1	Physical	Physical	Physical signaling (PHY)

Figure 1–2: ISO-OSI reference model and IEEE 802 standards model

The ZigBee Alliance is a consortium formed by several leading semiconductor and industrial manufacturers, distributors, and end users.

One of the tasks of this organization is the definition of the networking support and applications profiles that will use IEEE Std 802.15.4-compliant transceivers.

WIRELESS SENSOR NETWORKS

Wireless Sensor Networks (WSNs) are a subset of wireless networking applications focused on enabling connectivity without the use of wires to sensors and actuators in general.

Sensor networks can be classified according to the type of sensors, kind of application (industrial, medical, vehicular, etc.), environment where the network operates (explosive, vibration, acceleration, temperature, etc.), and by network parameters (network topology, required throughput, range, etc.). The IEEE 802.15.4 Working Group is chartered to focus on wireless sensor networks.

 Due to the length of the name "Wireless Sensor & Actuator Networks," the industry has adopted "Wireless Sensor Networks" instead. In any case, it is important to remember that the design of this type of network is meant to collect and to send information to wireless transceivers attached to a sensor and/or actuator.

The interest in WSNs has three faces. First, there is a need to lower the cost of sensor installation, which comes in the form of cabling, labor, materials, testing, and verification. For example, a limit-switch may cost less than a dollar, but the installation cost may range from $50 to $100. Furthermore, wiring regulations in the industrial and residential environment may require additional materials and activities for the wire installation, e.g., conduit and installation labor.

Second, cables require connectors that can get loose, lost, misconnected, or break (due to frequent access to neighboring devices). This problem is commonly known as "the last meter connectivity problem" and is named this way due to the analogous problem in the wide area networking called "the last mile connectivity problem."

The third aspect is that WSNs form the lower layer of intelligent maintenance systems enabling sensor-rich environments that generate abundant data that may be used to improve industrial operations. WSNs allow the gathering of more frequent data on a large number of machines and industrial systems in general. The use of large amounts of hardwired sensors networked to central systems brings a lot of complexity to the system, making it impractical in most cases.

A wireless solution for sensor networks provides flexible connectivity without the need for connectors. In addition, improved operator safety can be achieved by the use of wireless systems. Of course, WSNs share most of the issues surrounding wireless applications such as information security, authentication, small-scale radio-frequency propagation, antenna placement, and others.

Mobility is another benefit from the wireless solution, although in the WSN context, this capability is traded with "ease of installation." In other words, mobility is normally not a requirement for a WSN system, but certain mobility concepts can be used to enable ad-hoc networking. It is important to clarify that the term mobility in this context refers to relative motion of devices with respect to each other. For example, a wireless network within a moving machine is not considered mobile if the sensors are meant to communicate only with devices within that machine.

The set of advantages described is not enough to replace hardwired connections. The reliability and security (perceived and real) of wired networks can be higher than wireless communication systems. It is expected, however, that hybrid networks, wired and wireless, will coexist. Wireless sensors will act as extensions of wired networks wherever the wireless capability adds value to the specific application.

Challenges in the Design of Wireless Sensor Networks

The inertia slowing the widespread implementation of WSNs is the lack of standardized technologies that can address their requirements both at the application level and from the communications point of view. The focus of the wireless industry has been primarily on communications with higher data throughput, leaving short-range wireless connectivity behind.

An important feature required for WSNs is the capacity for easy installation of a large number of transceivers. This requirement itself has all of the issues surrounding wireless communication (e.g., range), plus other challenges unique to the application. An example of these challenges is the need to logically bind specific devices together—a lamp and its light switch.

Among the possible wireless communication technologies, we have the following:

- *Light-communications:* This includes technologies like the Infrared Data Association (IrDA®) standard among others. The main disadvantage of this technology is the need for an unobstructed line-of-sight among the devices operating in this type of network.

- *Inductive Fields:* This technology has been extensively used for Radio Frequency Identification (RFID) applications. The main disadvantages of this technology are very low range and high energy required by the network coordinator. In addition, field alignment may be required for efficient communications.
- *Ultrasound:* Similarly to inductive fields, ultrasound requires high energy from the network coordinator. Line-of-sight is not a major concern in this technology but form factor (miniaturization) is.
- *Radio-frequency (RF):* This technology does not require an unobstructed line-of-sight. The current state of the technology allows the implementation of low-power radio transceivers with both data-rates and ranges scalable according to the application.

RF technologies seem to provide a sufficient set of advantages for the implementation of wireless sensor networks, but they also have a set of challenges that need to be addressed. Some of the high-level issues in the design and implementation of WSNs are as follows.

Power Consumption

Some applications require the use of completely untethered RF transceivers (no access to external power); this implies the use of batteries or power scavenging. If batteries are used, they should last a long time, because the need for sensor maintenance due to battery replacement goes against the original intention of ease of installation and low-cost operation. A direct implication of limited power is a limited communication range.

> WSNs, in general, cannot make use of the "battery recharging" culture commonly associated with consumer electronic devices such as mobile phones and PDAs.

A common solution to overcome this obstacle is to use power cycling; that is lower the duty cycle of operation of the device (this solution brings to surface a synchronization problem, which can be solved with the appropriate networking procedures).

Some elementary calculations reveal that a AAA battery with capacity of 750 mAh powering off-the-shelf short-range radio transceivers (10 mA typical active current consumption) will last for two years if a duty cycle of less than 0.5% is maintained.

> The design of IEEE Std 802.15.4 allows the implementation of radio transceivers with extremely low power consumption in comparison to current wireless networking radio technologies.

Range

Due to governmental regulations and implementation economics, RF power outputs ranging from 0 dBm to 20 dBm are typical of wireless systems operating in unlicensed bands. The limited power establishes a limited connectivity range — the maximum distance between a transmitter-receiver pair is constrained.

In the context of WSNs, multihop network protocols are required to circumvent this problem; this in turn implies the need for suitable routing algorithms.

> An RF output power of 0 dBm with radios having a sensitivity of –70 dBm will have a range of 10 m (30 ft) for average indoor environments (using a log-distance path loss model with path loss exponent equal to 3).

Availability of Frequency Bands

The RF spectrum is a scarce resource regulated by most governments. However, there are special unlicensed bands, the use of which is allowed if the wireless devices operate within a set of rules that control the RF output envelope in time, frequency, and amplitude. In some bands, in some regulation regions, regulations stipulate that in order to transmit a larger RF output power, the energy may have to be spread — i.e., spread spectrum modulation should be employed. The use of unlicensed bands is free of charge, provided that the manufacturer can demonstrate conformance with the rules.

For WSNs, the most common bands used (or planned to be used) are:

- 868.0–868.6 MHz: Available in most European countries
- 902–928 MHz: Available in North America
- 2.40–2.48 GHz: Available in most countries worldwide
- 5.7–5.89 GHz: Available in most countries worldwide.

> Radio devices operating in unlicensed bands must comply with local regulations. In the U.S., the Federal Communications Commission (FCC) is the institution in charge of the regulation of these devices. Similarly, the European Telecommunications Standards Institute (ETSI) coordinates regulation efforts in Europe. Other countries around the world have their own regulatory agencies; many of them accept either FCC or ETSI acceptance as proof of compliance.
>
> IEEE Std 802.15.4 contains an informative annex focused on regulatory requirements worldwide. The material, although introductory in nature, provides the implementer with enough background information to initiate the certification endeavor.

Wireless networking standards operating in the global unlicensed bands offer potentially lower cost due to technology availability, high volumes, and limited or no reengineering (no re-tuning). Recently, the major focus of short-range wireless communication Original Equipment Manufacturers (OEMs) has been in the 2.4 GHz and 5.7 GHz bands due to the large bandwidth available.

The above remarks identify a coexistence challenge because incompatible technologies sharing the same band will be competing to gain and maintain access. Several studies have been conducted to evaluate the impact of this issue, revealing potential problems that depend on traffic load, spatial distribution, output power, transceiver density, and RF propagation parameters ([3], [9–12], and [27]).

IEEE Std 802.15.4 was designed to operate in the following bands:
868.0 to 868.6 MHz → 1 Channel (20 kb/s)
902.0 to 928.0 MHz → 10 Channels (40 kb/s)
2.40 to 2.48 GHz → 16 Channels (250 kb/s)
The first two bands combined are called the "low band," and the third is called the "high band".

Network Topology

To overcome the challenge of limited range, multihop network topologies are required. A special requirement for WSNs is the need for low maintenance. To achieve this goal, the network topology should be designed to allow low duty cycle operation among the member devices of the network. An analysis of network topologies and network layers suitable for WSNs is presented in Chapter 6.

At this point, it is important to note an often-overlooked fact: Wireless Networking implies bidirectional communications among the devices participating in the network. This characteristic in turn increases the reliability of the system and allows monitoring of the state of all the devices.

Self-Organization

To achieve ease of installation, it is required that the wireless network be self-organizing, that is, each sensor device start participating in a network without requiring special in-situ configuration (i.e., addressing, association, and traffic balancing).

Self-organization is a feature of ad hoc networks. Ad hoc wireless networks are a collection of transceivers that form a network without the aid of any fixed infrastructure or centralized administration. The network employs routing protocols to determine the proper message path from source to destination; for example, in one protocol, each radio keeps a list of adjacent radio links and updates a routing table when the network topology changes.

The implementation of a self-organization strategy is a mixture among network topology, security implementation and application requirements. This feature is commonly addressed by the definition of application profiles, which are not part of IEEE Std 802.15.4. Application profiles group together applications with common functionality to achieve interoperability among the devices participating in it.

Chapter 2 Low-Rate Wireless PAN Applications
...motivation for a vision

IEEE Std 802.15.4 was designed to be used in a wide variety of applications, requiring simple wireless communications over short-range distances with limited power and relaxed throughput needs. The applications that IEEE Std 802.15.4 addresses can be placed in the following classifications:

- *Stick-On Sensor:* These applications comprise wireless sensors that can operate completely untethered. This implies battery-powered transceivers. The focus of these applications is for monitoring or remote diagnostics.

- *Virtual Wire:* These are monitoring and control applications that can only be enabled through wireless connectivity, in places where a wired communications link cannot be implemented, e.g., tire pressure monitoring, motor bearing diagnosis, engine components, and so on.

- *Wireless Hub:* These are applications in which a centralized wireless bridge is added over a wired network. A wireless hub acts as a gateway between a wired network and a wireless LR-WPAN network. In many cases, the wireless hub network is formed with two transceivers, one being the hub and the other being an LR-WPAN device embedded in a PDA device.

- *Cable Replacement:* These applications attempt to add value through the removal of wires in consumer electronic portable devices. These applications have been addressed by another WPAN technology: IEEE Std 802.15.1 (Bluetooth). In this case, IEEE Std 802.15.4 offers a low power alternative (at least one order of magnitude lower) with the other benefits of reduced cost. Cable replacement differs from "Stick-On" sensor applications in that the former may use a continuous source of power (mains) or rechargeable batteries.

Figure 2–1 shows the LR-WPAN application scenarios just introduced.

It is important to point out that the focus of IEEE Std 802.15.4 is to be an application enabler, where the value will be in the application and not in the wireless capability. This vision comes at odds with others, where the application is the communications capability — e.g., cellular telephony.

Figure 2–1: Some LR-WPAN application scenarios

This chapter examines some of the IEEE Std 802.15.4 applications that several leading companies and visionaries have proposed and that have served as motivation for the creation of this standard.

 IEEE Std 802.11, IEEE Std 802.15.1, and IEEE Std 802.15.3 were created to target specific applications. In contrast, IEEE Std 802.15.4 was designed to address a wide range of applications in different market segments.

INDUSTRIAL AND COMMERCIAL CONTROL AND MONITORING — WIRELESS SENSORS

As previously introduced in Chapter 1, the interest in wireless connectivity for the industrial and commercial sector is centered on the need to lower the installation cost of sensors and actuators, while creating a sensing-rich environment as a lower layer for intelligent systems.

Initial implementations using IEEE Std 802.15.4 will most likely occur in monitoring applications with non-critical data, where longer latencies are acceptable. These industrial monitoring applications, in general, do not need high data throughput or constant updating. Instead, emphasis is placed on low power consumption in order to maximize the lifetime of the battery-powered devices that make up the network.

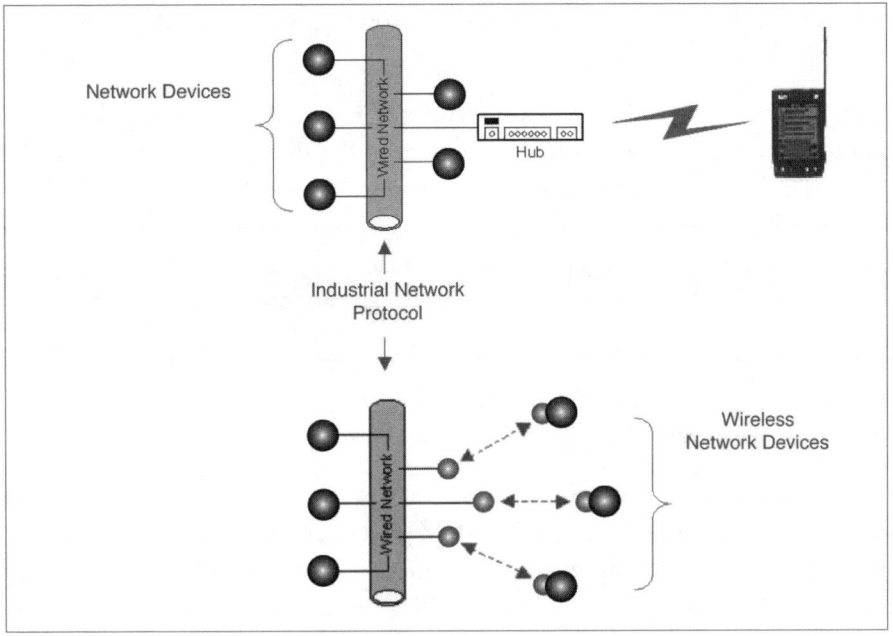

Figure 2–2: Typical industrial use cases

Typical applications in industrial automation will include wireless access points operating as gateways of a wired industrial protocol (e.g., DeviceNet™, Fieldbus®, INCOM™, and others). These access points will allow monitoring and parameter configuration of devices connected to the wired network from a PDA (wireless hub) or will allow a wireless link to a network device (cable replacement). Figure 2–2 shows the industrial uses cases explained above.

 In general, industrial applications do not have a particular network topology that characterizes them. The inherent network topology flexibility of IEEE Std 802.15.4 enables a wide range of applications in diverse industrial installations.

HOME AUTOMATION AND NETWORKING

The consumer and home automation market presents a significant potential because of its size. LR-WPAN devices will replace wires in home environment at a very low cost. Types of potential LR-WPAN devices include consumer electronic devices, personal computer peripherals, interactive toys and games, as well of home security, lighting control, and air conditioning systems [2]. Most of these applications have an industry group interested in using a low-cost, low data-rate wireless solution. Some of these industry groups are as follows: the Consumer Electronics Association (CEA), HomeRF (dismantled in January 2003), and the ZigBee Alliance.

Of particular interest is the analysis performed by the ZigBee Alliance that identifies home automation applications as providing the lower bound on required data throughput, and the upper bound on acceptable message latency, for LR-WPANs.

 IEEE Std 802.15.4 adopted a maximum data rate of 250 kb/s as a result of an analysis performed for home automation applications.

Consumer Electronics

The consumer electronics sector includes radios, televisions, VCRs, CD and DVD players, remote controls, and appliances in general. In this sector, the first application may be the truly universal wireless remote control. The opportunities to start bundling products and services with a common control mechanism (the IEEE Std 802.15.4-enabled remote) will open up many interesting opportunities. With the convergence of PC and consumer electronics markets, IEEE Std 802.15.4 gives the end user the ability to control devices from both segments with the same intuitive device they are familiar with: the remote control. It will now be possible to have remote control units assigned to individuals instead of being assigned to a piece of equipment (in addition to reducing the many remote control units in the home). Individualization and two-way remote controls become possible, to enable feature rich programming and control. However, an often-overlooked application is for communication links between devices to set volume levels, equalizer, and other control settings within a home entertainment center that may be distributed throughout the home.

PC Peripherals

PC peripherals, including wireless mice, keyboards, joysticks, PDAs, and games, represent a large segment inside the home for low-rate wireless connectivity. In addition, PC manufacturers and software companies are trying to change the interface to the PC, in order to transform it into a more consumer-oriented device. With the advent of remote controls and PDAs, this will become the way a PC is controlled. IEEE Std 802.15.4 devices will be able to provide the kind of low-power, low-cost solutions enabling this application scenario.

Home Automation

The home automation sector, which includes heating, ventilation, and air conditioning (HVAC), security, lighting, and the control of objects such as curtains, windows, doors, and locks, represents certainly a great opportunity for wireless innovation inside the home. In many cases, the thermostats in houses are located in the rooms where minimal time is spent (corridors, foyers, etc.). As a result, the temperature readings often do not correlate well with the temperature experienced by those living in the home, and hence, control of the environment in the home is often inefficient. Wireless thermostats could be dispersed throughout the home, and they could communicate with wireless vent controllers in the HVAC system to regulate the temperature in rooms individually. Window curtains could be drawn when the television is turned on during daylight hours; wall clocks could be coordinated with a master reference so that resetting after power failures would be automatic (no more blinking "12:00" on the VCR!). Home security could be as simple as the automotive remote keyless entry key fob—a single button press could lock all doors and windows; wireless smoke detectors and glass breakage sensors would connect to the home's security system. With the availability of all of these applications using the IEEE Std 802.15.4, a common remote control can be used to control all of these functions in addition to the consumer electronics devices.

Some manufacturers are planning to integrate IEEE Std 802.15.4 transceivers in light bulbs and ballasts. This would eliminate wiring (especially important for retrofit) and will enable standard wireless switches to be placed anywhere in the home without the need to break walls!

Home Security

Similarly to home automation, security sensors are dispersed around the home environment. These sensors are normally connected to a telephony-based backbone. Again, the value proposition of a wireless capability using LR-WPAN transceivers is based on the ease of installation.

There are two important challenges for wireless sensors in home security. The first one is related to power consumption of the sensor. Commonly, these sensors are not mains powered, because they are normally located on windows and doors. This challenge is overcome by IEEE Std 802.15.4, which as stated before, was designed with a simple yet efficient protocol stack. Furthermore, the medium access control (MAC) definition offers the upper layers an interface that allows full control to enable or disable the radio functionality. The second challenge is related to wireless range due to the propagation of radio frequency signals. The output power of IEEE Std 802.15.4 transceivers is bounded due to governmental regulations and by the need to conserve power (when necessary). IEEE Std 802.15.4 offers peer-to-peer protocol support that allows a network layer (not included in the standard) to implement a multihop network, effectively increasing the effective range.

Personal Healthcare

The personal healthcare sector includes sensors, monitors, and diagnostics. This is a field separate from medical telemetry; personal healthcare includes items such as pedometers and pulse rate monitors used by joggers and other athletes, and time and depth recorders used by scuba divers. These devices produce data that often need to be transferred to a display device; a wireless connection is often the only practical method to do this. Another application is health record maintenance; one's daily weight, temperature, and so on can be easily transmitted via a wireless link between the proper measuring device (scale, thermometer, etc.) and recorded on a personal computer or a PDA. With the availability of this information on the Internet, it is possible for healthcare professionals to access this data and monitor the status of persons remotely — no more need to drive to the weight loss center every other day!

Toys and Games

Low-rate wireless networks have much to offer in the toys and games sector, especially when the communication is among toys or a toy and a personal computer. Toys are extremely cost-sensitive; the limiting factor on such computationally intensive tasks as voice recognition and synthesis in a toy often is not the technical feasibility of the required task, but the need to perform it for a low total hardware cost. By adding a wireless link between the toy and a nearby personal computer, the cost of the toy may be reduced, because the toy only needs to include the wireless link and the necessary sensors and actuators (e.g., microphones and speakers). Such "PC-enhanced" toys may exhibit sophisticated behaviors (e.g., mobile robots), limited only by the capabilities of the computer and the wireless link. Other opportunities in this sector include wireless gaming between individuals or groups; in addition to simple wireless links between players, both toys and games could receive "updates" when within range of a computer, and then operate autonomously in the new mode, or with the new personality or feature, when leaving the range of the computer. Wireless gaming manufacturers using IEEE Std 802.15.4-based solutions will be able to address the installed base of games in addition to the new games, as the cost and power consumption of a wireless add-on module is targeted to be low enough to make it economically enticing to users to add on to their existing games.

A summary of the performance requirements of these application sectors is shown in Table 2–1. Note that the required data rate for most applications within each sector is significantly below the maximum rate shown.

Table 2–1: Requirements of home networking market segments

Sector	Max Required Data Rate	Maximum Acceptable Message Latency
Consumer Electronics	3 kb/s	16.7 ms
PC Peripherals	115.2 kb/s	16.7 ms
Home Automation	10 kb/s	100 ms
Personal Healthcare	10 kb/s	30 ms
Toys and Games	115.2 kb/s	16.7 - 100 ms

AUTOMOTIVE SENSING

Wireless communication is finding its way into cars, as driver comfort and the number of features offered increase. The first wave of wireless technology in a car came from the "remote keyless entry" application and its derivatives. The second wave is happening now in the form of "cable replacement" for telematics applications. Bluetooth has taken the lead in addressing this type of application, maintaining a major focus on telephony applications such as handling of a mobile phone from embedded devices in a car (in-car hands-free speech) and car personalization through data exchanged from a mobile phone to the car.

In general, cost plays a fundamental role for vehicle applications not related to safety and luxury convenience. Today, a car is loaded with a great amount of sensors and actuators distributed all around the vehicle. The use of these devices has originated a significant growth in wiring, causing a great impact on the installation costs, diagnostics, maintenance, and even fuel consumption!

The wireless option introduces flexibility in installation and an advanced alternative to wired connections. A special challenge for automotive applications is meeting the harsh automotive environment while maintaining the low-cost requirements.

An application example in the category of "virtual wire" is a tire pressure monitoring system. The system consists of four pressure sensors, mounted on each tire, and a central station to receive the collected data. Because the pressure sensors have to be mounted onto the tires, this application does not permit the use of any communication wires or power cables. Therefore, sensors have to be battery powered. Because it is impractical to replace the sensors or their batteries between tire changes, it is required that the sensor batteries last at least three years, and preferably five years. This puts significant constraints on the power consumption of the electronic components and requires power management capabilities. The data that needs to be communicated (the measured tire pressure) are, in most cases, only a few bits in size. This information is transmitted about every 1 to 10 minutes under non-alarm conditions. Unless there is a fast loss of pressure, the message latency is not of significant concern. In case of a sudden pressure loss, the central control unit should be notified immediately; in which case, the power consumption is not of concern because most likely the tire has to be replaced. Extreme automotive environmental conditions and the metallic structure of the car complicate the RF design. In addition, the shape of the rim has a significant impact on the radiation pattern from the wireless sensor. To overcome this issue, repeater devices, which should not add a significant cost to the system, can be added to the network to increase communications reliability. Of course, an implicit advantage not present in actual applications is the possibility to have two-way communications with the sensor, thus increasing the reliability of the system.

PRECISION AGRICULTURE

Another challenging application for LR-WPANs is precision agriculture, also known as precision farming. Precision agriculture is an environment-friendly system solution that optimizes product quality and quantity while minimizing cost, human intervention, and the variation caused by the unpredictability of nature. Today, agriculture is still both user and environmentally demanding. It is mainly hardware-oriented with manual and on-site control using independent dumb machines, which produce unpredictable quality and quantity. With the new paradigm of precision agriculture, farming would become more information and software-oriented, using automatic and remote-controlled, networked smart machines. This application requires large mesh-type networks consisting of potentially thousands of LR-WPAN devices linked with sensors. These sensors will gather field information such as soil moisture content, nitrogen concentration, and pH level. Weather sensors for measuring rainfall, temperature, humidity, and barometric pressure will also provide the farmer with valuable information. Each sensor will pass the measured data to its corresponding LR-WPAN device, which in turn will pass it through the network to a central collection device. In order for the sensor data to be useful, location-aware technology is necessary to correlate each sensor with its specific location in the field. The combined information will give farmers an early alert of potential problems and allow them to achieve higher crop yields.

The precision agriculture application is at the low end of the LR-WPAN application range, requiring the transmission of only a few bits of data per day by each deployed device. The data flow will be asynchronous in nature, with minimal restrictions on data latency. This combination of factors is advantageous for achieving long battery life. The challenge of this application is the topology of the network, because it requires a mesh topology where some devices serve as repeaters for others, relaying messages to the final destination, while still being power conscious to obtain the required usage life. The network should also be self-configuring because manual setup of a network of the proposed size is not feasible.

MISCELLANEOUS APPLICATIONS

A unique application scenario, falling within the consumer market, is a classroom calculator network. The teacher's workstation or the PAN coordinator would send tasks and math problems to each of the student's graphic calculators or the network devices. After completion, the students would upload their solutions back to the teacher's workstation. This network would need to support only a small number of devices, and it would require disallowing any peer-to-peer communication to prevent students from exchanging the solutions. The typical payload would be 100 to 500 bytes of information, sent several times per student per hour. It is desired that the batteries powering the calculator and communication function last the duration of a semester. Although this is definitely a higher throughput application than are the ones presented so far, it is still very well suited for an LR-WPAN.

Another important application in the category of "wireless hub" is remote metering and remote configuration. This type of applications enables improved hazard protection and operator convenience. Devices using an inexpensive LR-WPAN device will be able to exchange information wirelessly using PDAs, which can have previously programmed configuration information or contain software that will allow data to be synchronized with large databases.

Part II

Chapter 3   IEEE Std 802.15.4 Technical Overview

...warming up

The IEEE Std 802.15.4 Working Group established the goal of ultra-low complexity, low-cost, and extremely low-power wireless connectivity among inexpensive fixed, portable, and moving devices. With relaxed throughput and latency requirements, low-cost and low-power design could be achieved. The shared dream of "ubiquitous, untethered short-range communications" began to be realized.

Substantial costs can be associated with administration, sales, and marketing activities, as well as the cost of manufacturing and operation of wireless products. To achieve the goal of low total product cost as well as long battery life for low operating cost, the IEEE Std 802.15.4 provides reasonable trade-offs in several performance metrics. To control the administrative costs of both implementer and user, IEEE Std 802.15.4 devices employ unlicensed radio bands. Because IEEE Std 802.15.4 service is designed to be a short-range service without infrastructure, and able to support very large networks, both transmit and receive functions can and must consume little power and have a low operating cost. This chapter presents some of the unique features of IEEE Std 802.15.4 that enable this performance vision to be met.

How Low-Cost and Long Battery Life Are Achieved

Duty cycle

The battery is a relatively high-cost component in a transceiver system. The battery provides the energy for the communications, but at a cost of operation and replacement. Total system cost could not be minimized if a high-cost, exotic battery technology was required. The characteristics of low-cost batteries limit the choices for other attributes of the system. For example, the fully charged energy capacity and the instantaneous power delivery capacity of most batteries are interrelated. The fully charged energy capacity is achieved only if the instantaneous power delivery requirement is limited. Because of this limitation, to achieve long battery life, the energy must be taken continuously at an extremely low-rate, or in small amounts at a low-duty cycle. Due to the power consumption of practical wireless circuits, it is impractical to attain the desired battery life under constant operation; IEEE Std 802.15.4 was therefore designed to support very-low duty-cycle operation. The standard allows some devices to operate at low-duty cycles with both the transmitter and receiver inactive for over 99% of the time they are in operation.

However, in practical implementations there will always be a small amount of standby power consumed for timers, and so forth, when the wireless circuits are inactive. To reduce the time-average power consumption, both the active power and the standby power of an implementation should be reduced as much as possible. However, for a given technology and set of applications supported by the network, there is a practical limitation for both the active power and the standby power. For most applications, the active power is much larger than the standby power. Under this assumption, we can see that by reducing the duty cycle, low-power consumption levels, and associated long battery lifetimes can be reached.

For example, for a device with 10 mW active power and 10 μW standby power, if the duty cycle is 0.1%, then the time-average power drain is about 19.99 μW. If the device is supplied by a 750 mAh AAA battery, linearly regulated to 1 V, it will have a battery life of more than 37,000 hours, or more than four years.

To support low-duty cycles, the IEEE Std 802.15.4 beacon packet can be as short as 544 μs in the 2.4 GHz band, while the superframe period (the time between network beacons) may be extended from 15.36 ms to over four minutes. This results in a beacon duty cycle that may be set from 2.3% to 0.000216% (2.16 ppm). Further, the standard supports a non-beacon mode in which a network may operate without any beacons. The non-beacon mode enables the slaves in a master-slave star network, for example, to remain in standby mode indefinitely, only contacting the master (which may be mains powered, perhaps, and therefore capable of constant reception) when an event occurs. The slaves may therefore have an almost unlimited battery life, limited primarily by their standby power consumption. The non-beacon mode, in fact, may be a regulatory requirement for operation in the 868 MHz band, which has a maximum duty cycle limitation of 1%.

Modulation

With its low-cost and low-power consumption goals in mind, the IEEE Std 802.15.4 communications protocol was designed to support digital data communication only (i.e., analog service is not supported). Data-only service allows the modulation scheme to be chosen to be highly efficient, enabling a low-cost implementation. Likewise, the protocol supports only half-duplex operation so that the transmitter and receiver are not required to be active simultaneously.

Further, the data modulation schemes are simple raised-cosine-shaped Binary Phase Shift Keying (BPSK) in the 868/915 MHz PHY, and half-sine-shaped Offset Quadrature Phase Shift Keying (O-QPSK) in the 2.4 GHz PHY. Each of these forms of modulation maintains a peak-to-average carrier power ratio of one, which minimizes both power consumption and implementation complexity. Both modulation schemes further employ a spreading sequence to provide the benefits of Direct Sequence Spread Spectrum (DSSS) service; the 2.4 GHz PHY employs a form of multilevel orthogonal signaling, sending four bits per symbol, which simultaneously enables both a high-data rate (to return to standby mode quickly) and a relatively low-symbol rate (to minimize active power).

Use of Direct Sequence Spread Spectrum

DSSS is one of several techniques to increase the bandwidth of a transmitted signal. The wideband techniques provide improved communications qualities but usually sacrifice spectrum utilization. For instance, the conventional broadcast FM service employs a form of spectrum spreading (wideband FM) that, although it does not employ DSSS, exemplifies the improved Signal-to-Noise Ratio (SNR) communications benefits possible with wideband techniques.

DSSS can be modeled as applying a prearranged pseudo-random digital sequence to directly phase-modulate the already data-modulated carrier, at a rate in excess of the data rate. The resulting DSSS signal occupies a much greater bandwidth, albeit with a lower spectral power density. The signal is recovered by demodulating the received signal with a replica of the same modulating pseudo-random digital sequence. The replica is generated in the receiver by a technique that ensures that the replica is closely coherent with the modulation in the transmitted signal. An ideally recreated replica transforms the received DSSS signal back into a replica of the transmitter's original data-modulated carrier. The DSSS process spreads the original signal into a wider bandwidth for transmission over the channel, and then despreads the signal at the receiver to recover the original signal and the information it contained.

The benefits of the DSSS form of spectrum spreading are a direct result of the relative coherence of the pseudo-random digital sequences used to phase-modulate and recover the replica of the original data-modulated carrier. Only those signals that are modulated by the correspondent coherent pseudo-random digital sequence are despread back into a replica of the original data-modulated carrier. All other signals are further spread by the receiver despreading phase-modulation process.

Because a spreading process reduces the power spectral density of noncoherent signals being modulated, interfering signals, adjacent channel signals, and even other noncoherent spread-spectrum devices on the same channel have a lowered spectral density at the receiver within the narrow bandwidth of the original data-modulated carrier. All other signals are spread to a wider bandwidth, lower spectral density, and smaller energy contribution when filtered to the original data-modulated carrier bandwidth. This "processing gain" can be used, for example, to reduce requirements on channel filters, thereby lowering implementation costs.

The cost benefits of DSSS result from an SNR improvement of coherent versus noncoherent signals, but there are other pertinent effects also.

One effect is that the power spectral density of the signals emitted from the spread spectrum transmitter are lowered relative to the power spectral density of the unspread data-modulated carrier, and are less likely to interfere with narrowband services using the same band. This coexistence effect enables the use of multiple services in the same bandwidth. A DSSS service can be added to existing spectrum allocation with reduced interference to existing services and likewise be relatively unaffected also by those services. DSSS is a nearly ideal means to share spectrum.

A more subtle advantage of DSSS is that its implementations may be made of largely digital circuits, with few analog circuits. Because digital circuits follow Moore's Law, and shrink with improving integrated circuit lithography, they, and the resulting transceiver implementation, become cheaper to produce over time.

Finally, because of these favorable characteristics of spread spectrum operation, use of DSSS (or another form of spread spectrum, frequency hopping) is mandated by many regulatory bodies worldwide for the unlicensed bands under their jurisdiction. IEEE Std 802.15.4 makes use of these spectrum allocations to define a practical low-power, low-cost, unlicensed service.

 Moore's Law establishes that the number of transistors on an integrated circuit doubles every 18 months.

Transmitter power

Additionally, although IEEE Std 802.15.4 permits any legally acceptable output power, it requires only that the compliant device be capable of transmitting −3 dBm, well within the instantaneous power capacity of inexpensive battery sources, and within the capability of highly integrated and low-cost System-on-a-Chip (SoC) implementations.

Receiver sensitivity

The required minimum −92 dBm RF sensitivity specification in the 868/915 MHz PHY, and the −85 dBm RF sensitivity specification in the 2.4 GHz PHY, permit the use of a simple receiver to achieve compliance. Simple, low-cost receiver designs, with little radio-frequency amplification (which can be expensive in terms of power consumption) are possible.

Quality of service

To reduce implementation complexity, IEEE Std 802.15.4 does not support isochronous communication, and it does not support multiple classes of service within a single PAN. An exception is that the standard does support the use of Guaranteed Time Slots (GTS), which reserve network time for synchronous communication (avoiding potential channel access delays), at the discretion of the PAN coordinator. This extends the application space of IEEE Std 802.15.4 devices to include applications requiring low-latency communications, such as wireless joysticks and mice.

Although IEEE Std 802.15.4 is of unusually low complexity, the PHY layers support instantaneous link data rates of 20 kb/s, 40 kb/s, and 250 kb/s. The data rates are high relative to the data throughput in order to minimize device duty cycle.

In summary, IEEE Std 802.15.4 is designed to trade lower data throughput and higher message latencies for significantly lower cost and power consumption, while still providing useful communications for a wide class of services.

OTHER NOTABLE IEEE STD 802.15.4 FEATURES

Network components

An IEEE Std 802.15.4 network is composed of a set of IEEE Std 802.15.4 wireless devices. At least one network coordinator device and at least one network device form the network. Each network contains exactly one specialized network coordinator, called a PAN coordinator.

IEEE Std 802.15.4 defines two types of devices, the Full Function Device (FFD) and the Reduced Function Device (RFD). The FFD contains the complete set of MAC services and allows it to operate as a network coordinator or network device. The RFD contains a reduced set of the MAC services, and it can only operate as a network device.

RFDs were defined to allow the implementation of extremely simple devices that require minimal resources in terms of processing power and memory capacity. These characteristics have a direct impact on the implementation of lower cost devices to enable applications such as light switches, stick-on sensors, and others.

Multiple network topologies

An IEEE Std 802.15.4 network may be one of two basic topologies. The first topology is called a star topology, and it is formed around a full function device that is designated the PAN coordinator and that acts as a hub with a collection of additional full or reduced function devices that act as data terminal locations. In this regard, the PAN coordinator performs a special function in the network; it is the only device in the network that forms a direct link to more than one other device.

The second topology enables peer-to-peer communication without the direct involvement of a designated network coordinator as such (although a PAN coordinator is required somewhere in the network), and each device is able to form multiple direct links to other devices. The capability to form direct links with more than one other correspondent device ensures that the network graph can form redundant paths, but the complexity is increased. Consequently, many types of peer-to-peer network architectures can be supported.

One such complex network that can be supported by IEEE Std 802.15.4, but is not part of the specification, is the cluster-tree. A cluster-tree (see Figure 3–1) can be interpreted as a hierarchical tree of network device clusters. The cluster-tree arrangement provides complete connectivity, somewhat simplified routing, and a minimum number of direct links compared to a fully interconnected peer-to-peer structure, but it may result in increased data traffic latency.

Since they must do no message relaying, the leaves of a cluster tree (those devices at the bottom of the hierarchy) may be either FFDs or RFDs. All other devices in the network must be FFDs. Exactly one device in the network assumes the special role of the PAN coordinator, which has a function much like the root device of a hierarchical tree. The PAN coordinator may be selected because it has special computation capability, a bridging capability to other network protocols, or simply because it was among the first participants in the formation of the network. Other network devices may function as roots of individual clusters; these are designated as cluster heads, or coordinators. Surrounding each coordinator, a hierarchical tree may form in a typical parent–child relationship, but only one device in each cluster is the root or coordinator. The PAN coordinator may perform the coordinator function for its own cluster, but only a single device in the entire network functions as the PAN coordinator.

Networks topologies built on the IEEE Std 802.15.4 peer-to-peer communication feature are supported by the higher levels of the protocol stack which are outside the scope of IEEE Std 802.15.4. They are introduced here to show the flexibility of which the standard is capable; additional discussion is in Chapter 6.

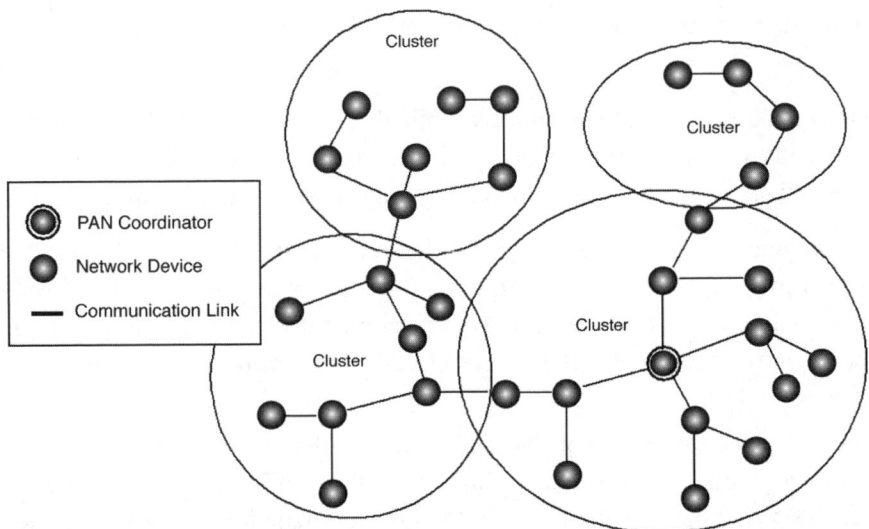

Figure 3–1:   Cluster-tree network

Channel Access

Regardless of the type of network employed, each network device employs a Carrier Sense Multiple Access-Collision Avoidance (CSMA-CA) protocol to avoid wasteful collisions when multiple simultaneous transmissions might otherwise occur. Exceptions to this are beacon transmissions, transmissions in GTSs, and acknowledgments, each of which is transmitted without the CSMA-CA process.

The CSMA-CA protocol is based on the shared nature of the RF channel. Whenever two or more senders are active on a channel simultaneously, the probability that any one of them is successful in sending a message decreases due to collisions and their mutual interference. In the RF environment, a great deal of the actual interference depends on the location of the competing transmitters and their respective correspondents, but the location information is often not available to the transmitting devices. One way to avoid collisions in the channel is to listen first and to transmit only if the channel is clear. The carrier sense ensures access to the clear channel and the possibility of a collision is reduced. In this fashion, the channel capacity is more fully utilized.

IEEE Std 802.15.4 specifies that the superframe begins with the beacon transmission, followed by the contention access period (CAP) and then the GTSs. The actual length of each is left as an upper layer decision for the PAN coordinator, but the CAP must be a minimum of 440 symbols in length, except for temporary reductions needed to accommodate additional beacon lengths for GTS maintenance.

The assignment of the GTSs is entirely at the discretion of the PAN coordinator, but the facility may be useful for two attributes associated with the use of contention-free assigned access, low latency, and bandwidth allocation.

- *Low latency* — For those applications that require minimum end-to-end delay for prioritized messages, slot assignment facilitates preemptive, out-of-order queue management in the protocols to minimize that delay. Examples might include alarm conditions relaying, or QoS differentiation based on application software.

- *Bandwidth allocation* — For those applications that may be less sensitive to delay, but generate a known data traffic rate, the facility to allocate bandwidth to each service permits the protocol to manage queue lengths at each network device. Traffic management is usually associated with network protocols above the MAC layer, but the capability must be established at the lower levels covered by IEEE Std 802.15.4 to be available to higher levels.

GTSs make higher level protocol services possible, but for those networks not implementing higher levels of service, the use of GTSs is optional.

Multiple PHYs

IEEE Std 802.15.4 employs two PHYs—one for the 868 and 915 MHz bands, and one for the 2.4 GHz band. Note that the 868/915 MHz PHY requires that a compliant device be capable of operating in both the 868 MHz band and the 915 MHz band. This stipulation reflects the desire to minimize the number of potentially incompatible products on the market, and it recognizes that there is little cost penalty in this requirement, due to the small frequency difference between the two bands.

The Task Group that developed IEEE Std 802.15.4 was comprised of volunteers from around the world, and therefore considered that compliant devices may be developed and used worldwide. To keep administrative costs low, IEEE Std 802.15.4 assumes that all devices will perform in unlicensed operation but each would fall under the provisions of the regulatory agency with jurisdiction for the particular service area. Unfortunately, the regulatory agencies worldwide have allocated the spectrum differently between and among themselves. There are three bands available for this type of service—one worldwide and two regional special cases.

For implementers, the choice of bands depends on more than simple technical considerations. The 868 and 915 MHz bands, although they are likely to be less crowded and may offer better QoS, are not available worldwide. The decision to implement products for a limited market, albeit with possible better service, falls in the realm of marketing and business rather than purely technical matters. There are "hidden" costs of distribution and marketing of otherwise identical products that differ only in frequency band. There are interference problems that bring up questions of how to keep products from moving from one regulatory region to another. Other applications (like wireless luggage tags) do not fit into a regional marketing strategy and do not present a consideration, but the possibility of regional strategies exists for some products.

Portions of the 2.4 GHz band are available nearly worldwide; the 868 MHz band is available in Europe, and portions of the 915 MHz band is available in North America, Australia, New Zealand, and some parts of South America. IEEE Std 802.15.4 defines a single channel with a raw data rate of 20 kb/s in the 868 MHz band. In the 902—928 MHz band, the standard defines 10 channels with a raw data rate of 40 kb/s (doubled chip rate with respect to the 868 MHz band). Higher in the spectrum, at 2.4 GHz, IEEE Std 802.15.4 defines 16 channels with a raw data rate of 250 kb/s.

Error control

IEEE Std 802.15.4 employs a simple full-handshake protocol to ensure reliable data transfer and good QoS. With the exception of broadcast frames (e.g., beacons) and the acknowledgment frame, each received frame may be acknowledged to assure the transmitting device that its message was, in fact, received. If a requested acknowledgment frame is not received by the transmitting device, the entire transmitted frame may be repeated.

To detect that a message has been received correctly, a cyclic redundancy check (CRC) is used. The message bits are treated as a long binary number and divided by a relatively large prime number. The quotient of the division is discarded, and the remainder of the division is transmitted along with the message. The same division with the same prime number is performed at the receiver, and a match signifies a high probability of an uncorrupted communication.

FOUR FRAME TYPES

There are four packet or frame structures, each designated as a PHY Protocol Data Unit (PPDU) in the standard for data transactions—a beacon frame, a data frame, an acknowledgment frame, and a MAC command frame. As shown in Figure 3–2, all frames are structured in a very similar fashion, with the primary differences in their purpose or payloads. Each PPDU is constructed with a Synchronization Header (SHR), a PHY Header (PHR), and a PHY Service Data Unit, composed of the MAC Payload Data Unit (MPDU) as a data structure that services the MAC protocol layer. The MPDU is constructed with a MAC Header (MHR), a MAC Footer (MFR), and a MAC Service Data Unit (MSDU)—excepting the MPDU of the acknowledgment frame, which does not contain an MSDU.

The MSDU is a data (i.e., payload) field component of a given frame containing information pertinent to the MAC services supported by the frame, including superframe identification and sequencing information, addressing information, and other information.

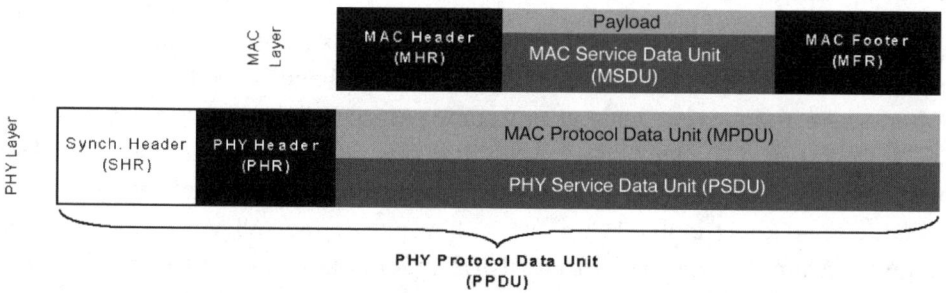

Figure 3–2: Frame structures format

Beacon frames

Transmission of the beacon frame is available only to FFD in the network, regardless of whether the network is a star, cluster-tree, or other topology. The beacon frame is provided as a service originating in the MAC protocol layer and interfaced to the PHY protocol layer. It has a number of uses, including superframe boundary marker, frame synchronizing signal, and association supervision, all as a service to the higher protocol layers.

As a superframe boundary marker, the beacon frame provides a timing reference to designate the boundaries of a superframe. In the higher level of the protocol, the superframe permits a fixed number of frames to be designated between beacon frames, whether each frame position is occupied or not.

As a frame synchronization signal, the beacon frame allows the synchronization of the superframe to a known starting time, preventing collisions, allowing receivers to sleep whenever there are no transactions, and permitting good message latency by limiting delay of traffic to the length of one superframe on any one link. Further, the synchronization allows the network device transmitting the beacon to become the timing reference for link communication and allows a relaxation of an individual network device's timing requirements. Each network device may have a less precise timing reference that is periodically adjusted to match the beacon device's reference.

Data frames

The data frame is available to all devices in any network, regardless of whether the network is a star, cluster-tree, or other topology. The data frame, like the optional beacon frame, is provided as a service originating in the MAC protocol layer and interfaced to the PHY protocol layer. It provides the primary data payload as a service to the higher protocol layers.

Acknowledgment frames

The acknowledgment frame is also available to all devices in any network, regardless of whether the network is a star, cluster-tree, or other topology. The acknowledgment likewise is provided as a service originating in the MAC protocol layer and interfaced to the PHY protocol layer. It provides only the acknowledgment for receipt of data as a service to the higher-level protocol layers for end-to-end message control. The acknowledgment frame does not contain a MAC payload.

MAC command frame

The MAC command frame is also available to all devices in any network, regardless of whether the network is a star, cluster-tree, or other topology. The MAC command frame, like the other frames, is provided as a service originating in the MAC protocol layer and interfaced to the PHY protocol layer. It provides the primary supervisory payload as a service to the MAC protocol layer.

For each of these four frame types, the MPDU is further encapsulated with an additional preamble and PHY header to form the physical layer service data unit (PSDU). This complete data structure allows a receiver to achieve symbol synchronization with the frame and obtain the frame length information for checking purposes.

Chapter 4 Physical Layer

...from bytes to watts

The physical layer (PHY) provides the interface with the physical medium where the actual communications occurs. The PHY layer is the lowest component in the ISO/OSI model and is in charge of providing control (activation and deactivation) of the radio transceiver, energy detection, link quality, clear channel assessment, channel selection, and the transmission and reception of message packets through the physical medium [26].

REGULATORY CONSIDERATIONS

Governments around the world regulate and administer the radio-frequency spectrum. In most cases, there are band allocations for unlicensed operation, given that the manufacturer can ensure operation within some pre-established limits in output power, duty cycle, modulation, and other parameters. Station licenses, along with an identification call sign, are issued to transmitters for many of the allowed services, but there is also a provision for the operation of some devices that do not require station licenses. The particulars of unlicensed services vary throughout the world, but generally, the devices must conform to a set of regulatory limitations that are specific for each operation. Transmitters for each operation must pass a testing protocol to ensure that they meet the specific requirements. Once a particular design has passed the regimen of testing, it is issued a certification or type approval for that service. IEEE Std 802.15.4 devices are targeted toward type approval.

Implementations of IEEE Std 802.15.4 will have to conform to local regulations of the country where it will operate. In the case of Europe, Japan, Canada, and the United States the regulation consists in unlicensed but type-approved DSSS service. It is the constraint for unlicensed, but type-approved compliance that dictates the frequency bands of operation, as well as a few of the other characteristics of the service.

IEEE Std 802.15.4 is written so that conforming devices can be manufactured to operate in any of three particular bands. Two of these bands are limited to specific geographical regions, but one band is available for nearly worldwide service. Details of these bands are presented next.

European Regional Case

In Europe, the European Telecommunications Standards Institute (ETSI) has published recommendations that are recognized generally by all regulatory agencies, but each service area falls under individual national type-approval authority. Within the European service area, there is one common band with operation allowed between 868.0 and 868.6 MHz that supports a single channel of low data rate service with less than 1% transmission duty cycle [4], [5]. Again, this band is unique to European service, and regulations elsewhere preclude operation of unlicensed devices in this band.

IEEE Std 802.15.4 868/915 MHz PHY requires that a compliant device be capable of operation on both the 868 and 915 MHz bands. For the purposes of the standard, they are considered to be a single, contiguous band.

In May 2002, the Electronic Communications Committee (ECC) within the European Conference of Postal and Telecommunications Administrations (CEPT) released their strategic plan for the future use of the frequency band 862–870 MHz. This plan delineates a roadmap to increase the frequency band available for spread spectrum devices. The allocated band would operate in the 865–868 MHz band, allowing the addition of at least three radio channels to a future release of IEEE Std 802.15.4.

North American Regional Case

Within the United States, the national regulatory agency is the Federal Communications Commission (FCC). The FCC has specific authority only within the United States; however, the FCC regulations are used as a model by many other nations in the Americas and the Pacific Rim. Just as in Europe, there is a unique band available for service; this band is called the 915 MHz band and it covers the range between 902 and 928 MHz. This band enables the provision of ten channels of low data rate service. With some exceptions, unlicensed operation in this band is unique to North America, and regulations elsewhere preclude operation of unlicensed devices in this band.

Worldwide Case

To obtain economies of scale in product design, marketing, and distribution, and to enable applications that may require roaming between different regulatory regions, it is desirable to employ a single band that is available on a nearly worldwide basis. The ideal band would be unlicensed, have sufficient width to enable the use of many channels, and be high enough in the spectrum so that relatively efficient antennas are possible without being so high that single-chip implementations employing low-cost integrated circuit processes are precluded.

The band selected as best meeting these requirements is the 2.4 GHz Industrial, Scientific, and Medical (ISM) band, which extends from 2400 to 2483.5 MHz. This band, with very few exceptions, is available worldwide without licensing [6], [22], [23] and, with a wavelength of 12.25 cm, enables reasonably efficient yet physically small antennas. Further, it is compatible with modern silicon integrated circuit processes, and the wider bandwidth available enables the provision of sixteen channels of higher data rate service.

IEEE Std 802.15.4 was designed for regulatory compliance in each of these bands. For example, in the 868 MHz band, the product duty cycle is limited; the non-beacon mode of IEEE Std 802.15.4 may be used to minimize the duty cycle of devices in this band. In many cases, some type of spread spectrum operation is required for operation in unlicensed bands. The use of DSSS in IEEE Std 802.15.4 allows this regulatory requirement to be met, while also enabling a low-cost product implementation with good transmission range.

FREQUENCY BANDS AND DATA RATES

In response to the regulatory availability of the three bands for unlicensed operation, IEEE Std 802.15.4 specifies technical objectives for operation within each band:

- *868–868.6 MHz Band:* This unlicensed band is available in most European countries for a 20 kb/s DSSS service. IEEE Std 802.15.4 also refers to this band as the "868 MHz" band.

- *902–928 MHz Band:* Some portions of this unlicensed band are available in North America, Australia, New Zealand, and some countries in South America for 40 kb/s DSSS service. IEEE Std 802.15.4 also refers to this band as the "915 MHz" band (which represents the middle frequency of the band).

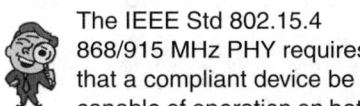

The IEEE Std 802.15.4 868/915 MHz PHY requires that a compliant device be capable of operation on both the 868 and 915 MHz bands. For the purposes of the standard, they are considered to be a single, contiguous band.

- *2.4000–2.4835 GHz Band:* This third unlicensed band is available in most countries worldwide for the faster 250 kb/s DSSS service. This band is referred to as the 2.4 GHz band.

Due to its nearly world-wide availability, the 2.4 GHz PHY may be the first choice for many IEEE Std 802.15.4 applications, especially those involving products, such as wireless luggage tags, that involve travel between regulatory regions. Even for non-mobile applications, the 2.4 GHz band offers advantages of scale, distribution, and marketing: A single product can be sold in multiple locations around the globe, without concern for the regulatory region in which it will be used. This drives down production costs, while eliminating the supply chain expense of tracking multiple products to multiple destinations.

However, for these same reasons, the 2.4 GHz band is used by many other services, from microwave ovens to WPANs and WLANs. This can result in congestion that is unacceptable for some applications and markets. To address this issue, the IEEE Std 802.15.4 868/915 MHz bands are available as an alternative. This feature allows system designers to use these regional bands in order to avoid the potentially crowded 2.4 GHz band. The regional bands may be a good design choice for applications such as utility meter reading that are inherently regional in nature and have limited device mobility. Similarly, diverse sensors in different industrial segments can make use of this feature.

Data Rates

Due to the physical characteristics of each band and the regulations where they are used, IEEE Std 802.15.4 specifies different data rates and modulations for the three bands used in the two PHYs.

Table 4–1 shows the data rates (bit and symbol) and the direct sequence spread spectrum parameters specified in each band.

A more detailed explanation of the modulation parameters will be introduced in the band specification sections below.

Table 4–1: IEEE Std 802.15.4 frequency band and modulation parameters

Band	Frequency Band	Bit Rate	Symbol Rate	DSSS Spreading Parameters	
				Modulation	Chip Rate
868 MHz	868–868.6 MHz	20 kb/s	20 ksymbols/s	Binary Phase Shift Keying (BPSK)	300 kchip/s
915 MHz	902–928 MHz	40 kb/s	40 ksymbols/s	Binary Phase Shift Keying (BPSK)	600 kchip/s
2.4 GHz	2.4–2.4835 GHz	250 kb/s	62.5 ksymbols/s	Offset Quadrature Phase Shift Keying (O-QPSK)	2 Mchip/s

CHANNEL ASSIGNMENT

IEEE Std 802.15.4 specifies a total of 27 channels across the three frequency bands. The channels are numbered 0 to 26; one channel is available in the 868 MHz band, 10 channels in the 915 MHz band, and 16 channels in the 2.4 GHz band. The center frequencies of these channels are shown in Table 4–2.

Table 4-2: IEEE Std 802.15.4 channel assignment

Band	Channel	Center Frequency (MHz)	Availability
868 MHz Band	0	868.3	Europe
915 MHz Band	1	906	Americas
	2	908	
	3	910	
	4	912	
	5	914	
	6	916	
	7	918	
	8	920	
	9	922	
	10	924	
2.4 GHz Band	11	2405	Worldwide
	12	2410	
	13	2415	
	14	2420	
	15	2425	
	16	2430	
	17	2435	
	18	2440	
	19	2445	
	20	2450	
	21	2455	
	22	2460	
	23	2465	
	24	2470	
	25	2475	
	26	2480	

PHY BIT LEVEL COMMUNICATION

The PHY level of the IEEE Std 802.15.4 protocol is responsible for the establishment of the RF link between two devices. The PHY is also responsible for bit modulation, demodulation, and synchronization between the transmitter and receiver. Finally, the PHY also is responsible for packet level synchronization.

IEEE Std 802.15.4 specifies three different data rates in each of the three different bands. The two lower rates, used in the 868/915 MHz PHY layer, are specified using a BPSK modulation scheme, and the highest rate, used in the 2.4 GHz PHY layer, is specified using the O-QPSK modulation scheme with an m-ary quasi-orthogonal modulation technique.

2.4 GHz Band Specification

The 16-ary quasi-orthogonal modulation technique specified in the IEEE Std 802.15.4 2.4 GHz PHY layer is a method of data modulation that utilizes a particular 32-chip, pseudo-random sequence to represent four bits and simultaneously accomplish the spreading modulation. The data modulation is performed by means of cyclic rotation and/or conjugation (inversion of chips with odd indices) of the sequence. The pseudo-random sequence is started in different places, depending on the modulating data, transmitting four bits in each symbol period.

While five bits could be transmitted by the choice of 32 chips, four were chosen for the 2.4 GHz PHY to minimize implementation complexity. The transmitted 32-chip pseudo-random sequence is allowed to start only at every fourth chip of the sequence. Symbols 0–7 represent cyclic shifts in multiples of four chips. Symbols 8–15 use the same shifts as symbols 0–7, respectively, but use the conjugated sequence (i.e., the odd-indexed chips are inverted).

The IEEE Std 802.15.4 2.4 GHz PHY layer specifies a symbol rate of 62.5 ksymbols per second with four bits in each symbol; therefore, 250 kb/s service is attained. The 32-chip pseudo-random sequence to be transmitted is split between the orthogonal I and Q channels of the O-QPSK modulator, with the even-indexed chips placed on the I channel and the odd-indexed chips placed on the Q channel. A one-half chip delay is placed in the Q channel, creating the offset for O-QPSK. Because 32 (now complex) chips are transmitted in one symbol time (16 µs), the overall chip rate is 2 Mc/s. The chip rate in either I or Q channel, however, is 1 Mc/s.

The bit level processing consists of assembling four bits into a symbol, converting that symbol to a cyclically-rotated 32-chip sequence as shown in Table 4–3, and modulating that chip sequence on the I or Q channel, respectively. The process is diagrammed in Figure 4–1.

The resulting combination of complex modulation, including the offset between I and Q channels is shown in Figure 4–2.

 Initially two distinct pseudo-noise (PN) sequences were considered for the I and Q channels of the quasi-orthogonal modulation in the O-QPSK, but the two were later combined to use a single PN sequence separated into even and odd bits between the I and Q channels. The advantage of a single PN sequence lies in that a single correlator is required that can be shared between the I and Q channels, thereby reducing implementation complexity.

Table 4–3: Symbol to chip mapping

Data Symbol (decimal)	Data Symbol (binary) (b3 b2 b1 b0)	Chip Values (c0 c1... C30 C31)
0	0000	11011001110000110101001000101110
1	0001	11101101100111000011010100100010
2	0010	00101110110110011100001101010010
3	0011	00100010111011011001110000110101
4	0100	01010010001011101101100111000011
5	0101	00110101001000101110110110011100
6	0110	11000011010100100010111011011001
7	0111	10011100001101010010001011101101
8	1000	10001100100101100000011101111011
9	1001	10111000110010010110000001110111
10	1010	01111011100011001001011000000111
11	1011	01110111101110001100100101100000
12	1100	00000111011110111000110010010110
13	1101	01100000011101111011100011001001
14	1110	10010110000001110111101110001100
15	1111	11001001011000000111011110111000

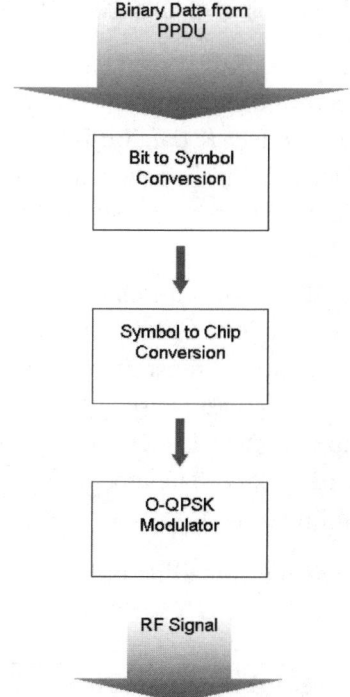

Figure 4–1: 2.4 GHz modulation and spreading

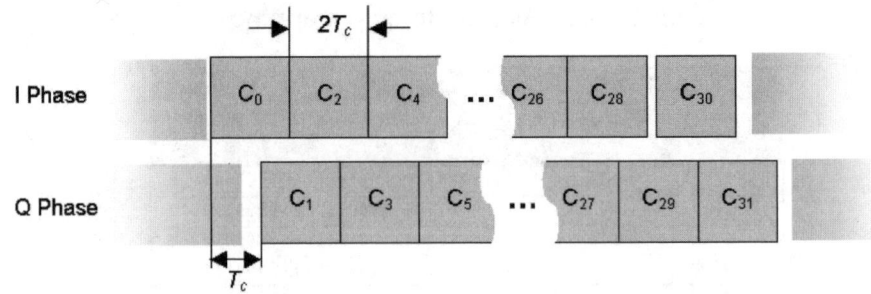

Figure 4–2: O-QPSK chip offsets

Because 32 chips are transmitted in one symbol time (16 µs), the overall chip rate is 2 Mc/s, and the length of a chip is Tc = 0.5 µs. Successive chips in each of the I and Q channels start every 2 Tc.

868/915 MHz Band Specification

The 868/915 MHz PHY layer specifies DSSS employing BPSK, with a data rate of 20 kb/s in the 868 MHz band and 40 kb/s in the 915 MHz band. A complex signal path is not required.

The 868/915 MHz PHY specifies differential encoding of the transmitted data bits. If the raw data bit is "0," the BPSK data bit is transmitted in the same phase as the previous BPSK data bit, while if the raw data bit is "1," the BPSK data bit is transmitted in phase opposite to the previous BPSK bit.

The 868/915 MHz PHY specifies "conventional" DSSS, in which a single, 15-chip, pseudo-random sequence is transmitted in a symbol period to represent a "1," and the inverse of the sequence is transmitted to represent a "0." This process is illustrated in Figure 4–3. The chip rate is specified as 300 kc/s in the 868 MHz band, for a data rate of 20 kb/s; in the 915 MHz band, the specified chip rate of 600 kc/s enables a data rate of 40 kb/s. The reception process used to recover the transmitted data is shown in Figure 4–4.

The 868/915 MHz modulation and spreading process is illustrated in Figure 4–5.

The BPSK modulation utilizes a different PN sequence, but it shares that sequence between the two lower data rates. The difference in data rates is due to the difference in chip rates, but the number of chip bits per data bit and the bit pattern are the same between the two.

Figure 4–3: DSSS modulation

Figure 4–4: DSSS demodulation

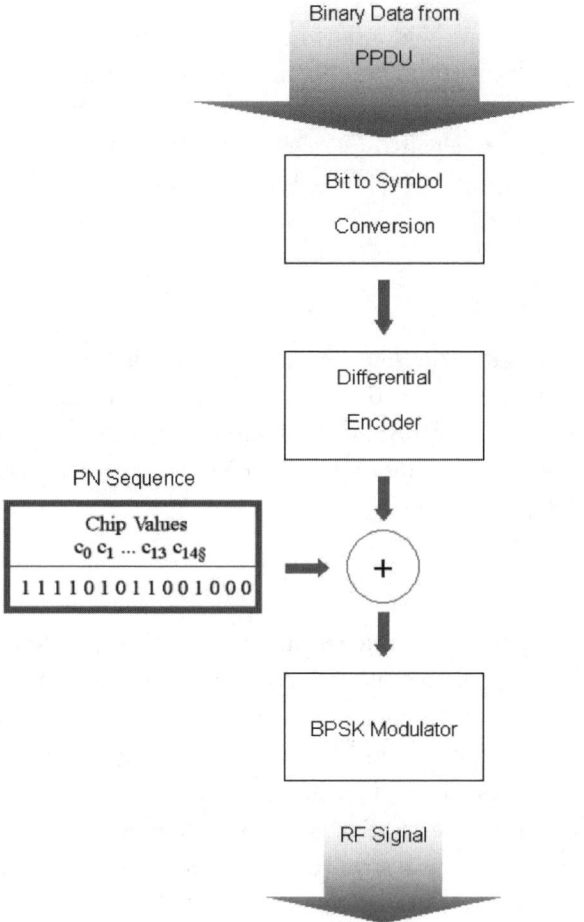

Figure 4–5:   868/915 MHz modulation and spreading

RADIO CHARACTERISTICS

IEEE Std 802.15.4 radio specification was designed to allow the implementation of low-cost digital integrated circuit designs. Most of the technical radio specification can be considered relaxed with respect to other radio technologies. The following paragraphs present some of the characteristics of the IEEE Std 802.15.4 radio.

Power Outputs

IEEE Std 802.15.4 provides for a wide transmitter output power range, but the device must be capable of transmitting −3 dBm. The upper limit of power output is designated by the regulatory agency associated with the local use. For example, in the United States, some services using DSSS in the 2.4 GHz band are allowed up to 1 Watt of transmitter power [22], however, in Europe the limit is 100 milliwatts in the same band [6].

Sensitivity

IEEE Std 802.15.4 specifies that the receiver must be capable of correctly decoding a signal with an input power of −85 dBm or less in the 2.4 GHz band. In the lower frequency bands where BPSK is used, the receiver must be capable of correctly decoding a signal with an input power of −92 dBm or less. Better sensitivity is not prohibited.

Range

In free space, the path loss between a transmitter and a receiver is dependent solely on the distance between them, and is independent of the frequency used. There is therefore no inherent difference in range between the IEEE Std 802.15.4 frequency bands. The type of antenna used, however, can give the illusion of a frequency-dependent path loss: A constant-gain antenna, such as a half-wave dipole, has an effective aperture (area) that decreases as the frequency of operation increases. (The dipole gets physically smaller as the frequency increases, leading to a reduction of its effective area.) At higher frequencies, the dipole therefore intercepts less of the transmitted radiation and produces less received signal at its terminals. Users of a communication link employing dipole antennas on both transmitter and receiver would therefore notice a range reduction as the frequency of operation increased, although the effect was due to the antennas employed, rather than an increase in path loss.

 Antennas that have a constant effective aperture exist; such antennas have gain that increases as the frequency of operation increases. An example of such an antenna is the parabolic dish antenna. Users of a communication link employing parabolic dish antennas of fixed dimensions on both transmitter and receiver would notice a range extension as the frequency of operation was increased—just the opposite effect of that seen by the users employing dipole antennas.

The above notwithstanding, many implementations of the IEEE Std 802.15.4 are expected to employ constant gain (dipole) antennas. For these implementations, and in the case of free space, the ratio of the power at a receiving antenna to the power at a transmitting antenna is given by the Friis model:

$$\frac{P_R}{P_T} = G_T G_R \left(\frac{\lambda}{4\pi d}\right)^2 = G_T G_R \left(\frac{c}{4\pi f d}\right)^2$$

where P_R and P_T are the power values at the receiving antenna and transmitting antenna, respectively, (in watts); G_R and G_T are the power gains of the receiving antenna and transmitting antenna, respectively; λ is the wavelength (in meters); f is the frequency (in Hertz); and c is the speed of light (in meters/second).

This same equation can be expressed as a basic path loss L_B in decibel form, with the appropriate substitutions for the constants as:

$$L_B(dB) = 32.44 + 20\log_{10} f_{MHz} + 20\log_{10} d_{km}.$$

For simplification, we treat the 868 and 915 MHz bands as approximately 1 GHz and the 2.4 GHz band as approximately 2 GHz so that the losses can be expressed as:

$$L_{B\,@\,1GHz}(dB) \approx 92 + 20\log_{10} d_{km}$$

$$L_{B\,@\,2GHz}(dB) \approx 92 + 20\log_{10} d_{km}.$$

Because lower frequency bands must exhibit a sensitivity of at least –92 dBm, for a transmitter rated at 0 dBm, and path loss of 92 dB, the maximum free space range is approximately 1 km ($\log_{10}[1]=0$).

Beware that the above derivations are using the free-space radio propagation model. These calculations are ideal; make sure you finish reading this section where a real-life model is used.

For the higher 2.4 GHz band, the receiver must exhibit a sensitivity of –85 dBm; for a transmitter rated at 0 dBm, similar calculations will show a maximum free space range of approximately 220 meters.

These ranges are for free space, isotropic antennas, and a perfect power match, without interference, and represent the maximum theoretical distances. The free-space model does not consider several environment parameters that impair the RF propagation, including wave reflection, diffraction, and scattering. A commonly used model that approximates the propagation behavior in RF channels in real environments is the log-normal shadowing model. Figure 4–6 illustrates a typical indoor propagation effect and the variance in range obtained due to the effect of shadowing (transmitting a message with an output power of 0 dBm). As it can be seen, the distance covered is a function of the frequency used. Similarly, the variability shown is a function of the characteristic of the wireless channel. The path loss coefficient in indoor environments can vary between $n = 2$ and $n = 4$ typically. Furthermore, the indoor variance in the received power ranges from $\sigma = 3$ to $\sigma = 11$.

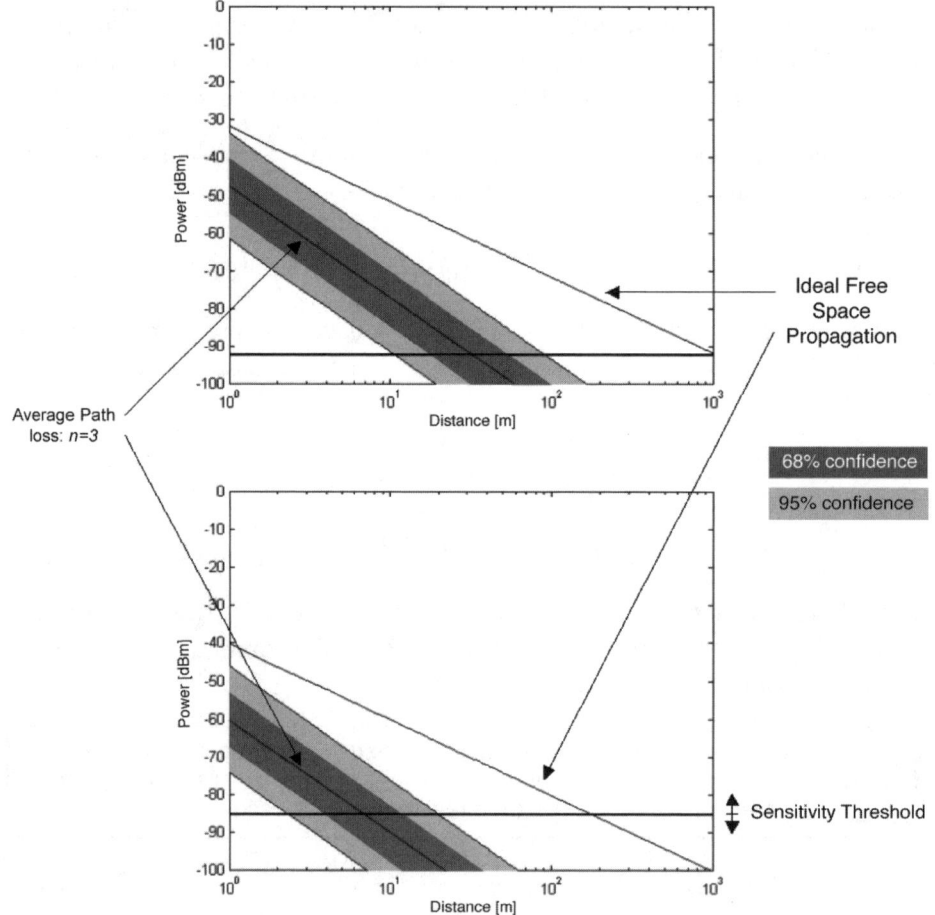

Figure 4–6: Typical indoor range (0 dBm transmission) with normal distribution ($\sigma = 7$dB): (a) 915 MHz, (b) 2.4 GHz

Receiver Selectivity

The DSSS signal is, by definition, a wideband variation of the modulated data signal. The DSSS spreading provides the communications benefits already discussed, and it provides a relaxed RF selectivity requirement. In addition, IEEE Std 802.15.4 channels are widely spaced (5 MHz in the 2.4 GHz band) relative to their own bandwidth (3 MHz null-to-null; about 1.5 MHz noise bandwidth in the 2.4 GHz band). With the 0 dB adjacent channel requirement, as well as the relaxed requirement for channels further away, little selectivity is needed by the implementer to achieve the requirements.

Channel Selectivity and Blocking

IEEE Std 802.15.4 specifies an adjacent channel rejection for service in the 902–928 MHz band, where 10 channels are allowed, and in the 2.4 GHz band, where 16 channels are allowed. In the 868 MHz band only one channel exists and an adjacent channel specification is not meaningful. For the 902–928 MHz and 2.4 GHz bands, the receiver must reject an interfering adjacent channel signal that is at the same level (0dB difference) as a simultaneous on-channel signal. One interfering signal at a time is specified at that level.

In addition, IEEE Std 802.15.4 also specifies an "alternate" channel rejection for service in the 902–928 MHz and 2.4 GHz bands. The alternate channel is the next nearest channel to the adjacent channel, or two channels away from the channel of operation. The receiver must reject an interfering alternate channel signal that is at a level 30 dB higher than a simultaneous on-channel signal. One interfering signal at a time is specified at that level.

The specification of these interference levels ensures reliable communication in the presence of multiple colocated WPANs, each on a different channel. The specification of a single interference signal reflects the relatively low traffic envisioned for the service.

To ensure that the receiver does not behave badly in other strong signal conditions, the specification provides a maximum acceptable input level that must not cause an excessive error rate. The IEEE Std 802.15.4 receiver must withstand a signal input of at least –20 dBm without unacceptable error rates.

There is no intermodulation specification in either IEEE Std 802.15.4 PHY layer.

PHY SERVICES

The PHY layer provides an interface between the physical radio channel and the MAC sublayer, through the use of two services. These services are the PHY data service and the PHY management service (called the PHY layer management entity or PLME) and are accessed by the PHY layer data service access point (PD-SAP)

The services of the PHY layer are the capabilities it offers to the MAC sublayer.

and the PHY layer management entity service access point (PLME-SAP), respectively.

> There are four types of service primitives. They are as follows:
>
> - *Request:* The request primitive is passed from the calling layer (also called the user layer) to request that a service is initiated.
>
> - *Indication:* The indication primitive is passed from the service layer to the user layer to indicate an internal event of significance. This event may be logically related to a remote service request, or it may be caused by a service layer internal event.
>
> - *Response:* The response primitive is passed from the user layer to the service layer to complete a procedure previously invoked by an indication primitive.
>
> - *Confirm:* The confirm primitive is passed from the service layer to the user layer to convey the results of one or more associated previous service requests.

PHY Data Services

The PHY data service provides three primitives to the MAC sublayer: data-request, confirm, and indication as shown in Figure 4–7. The asynchronous design of this service does not require the use of a data-response primitive.

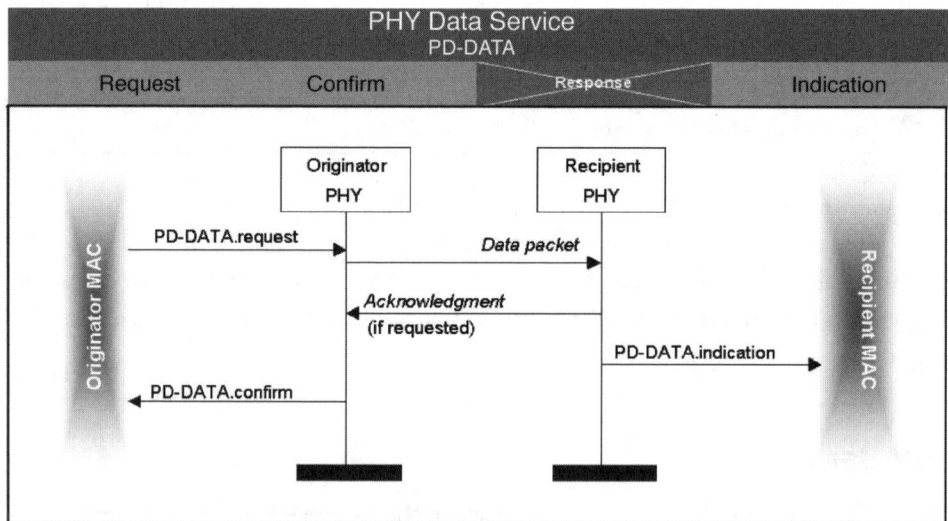

Figure 4–7: Message sequence diagram for data packet exchange mechanism

PHY Management Services

The PHY management service provides support for commands to control communication settings and the radio control functionality. The PLME primitives are summarized in Figure 4–8. The following paragraphs provide an overview of these primitives.

Primitive	Category	Description	Request	Confirm	Response	Indication
GET	Communication Settings	PHY PAN information base management	x	x		
SET			x	x		
SET-TRX-STATE	Radio Control	Enables/Disables radio system	x	x		
CCA	RF Energy Sensing	RF energy sensing: Clear Channel Assesment	x	x		
ED		Energy Detection	x	x		

Figure 4–8: PHY management service primitives

PHY PAN Information Base Management Primitives

The PHY PAN Information Base (PIB) contains configurable attributes to manage the PHY layer. These attributes can be read or written by the PLME-GET and PLME-SET primitives.

Figure 4–9 shows the message sequence diagrams for the procedure of reading or writing the PIB attributes.

Figure 4–9: Message sequence diagram for PHY PIB reading and writing mechanism

ENABLING AND DISABLING THE PHY

The radio receiver can be enabled or disabled by means of the PLME-SET-TRX-STATE primitive. The purpose of this primitive is to control the radio transceiver and enable lower power consumption. Figure 4–10 shows the message sequence diagram for this primitive.

Figure 4–10:   Message sequence diagram for transceiver enabling mechanism

CLEAR CHANNEL ASSESSMENT

Before the transmission of packets in a non–beacon-enabled network or in the contention access period of a beacon-enabled network, the MAC instructs the PHY to perform a clear channel assessment (CCA) before sending data and MAC command frames.

When the clear channel assessment is requested, the PHY layer enables the receiver, performs a CCA measurement, and then disables the receiver. Once the clear channel assessment measurement is completed, the PHY layer issues an PLME-CCA.confirm indicating if the channel is busy or not. Figure 4–11 illustrates the CCA mechanism using the PLME-CCA primitives. The figure also shows the message sequence chart for the clear channel assessment mechanism.

Figure 4–11: Message sequence diagram for the clear channel assessment mechanism

ENERGY DETECTION

The PLME-ED primitive allows a device to perform RF energy detection in the actual channel where it is operating. The measurement performed is similar to the one carried in PLME-CCA, but with higher resolution, it returns an energy level that ranges from 0 to 255. The use of this primitive can enhance the functionality of the network layers. Figure 4–12 shows the message sequence diagram for the energy detection process.

PHY PACKET STRUCTURE

The PHY Protocol Data Unit (PPDU) is the packet data structure at the PHY protocol level that modulates the wireless transmitter. The PPDU encapsulates all data structures from higher levels of protocol. The PPDU consists of three components: first, a synchronization header; second, a PHY header; and third, a variable length payload containing the PHY layer service data unit.

PPDU synchronization header — The PPDU header consists of two fields, a preamble and a start-of-frame delimiter. The preamble consists of 32 bits, all set to binary zero (recall that a binary zero gets encoded in a chip pattern). The preamble field allows a receiver a sufficient number of bits to achieve chip synchronization and bit synchronization. The start-of-frame delimiter consists of the 8-bit pattern

"0xe6" (11100101), and allows the receiver to establish the beginning of the packet in the stream of bits.

PHY header — The PHY header is a single 8-bit field with the MSB reserved and the remaining low-order bits used to designate frame length information. Packet lengths of 0 to 4 and 6 to 7 bytes are reserved. Packets of length 5 bytes are MPDU acknowledgment packets, and packets with 9 or more bytes are MPDU payloads for the MAC protocol layer service.

PHY payload — The PHY payload is composed of only one field called the physical layer service data unit (PSDU). The PSDU is variable length in nature and carries the data payload of the PPDU. All packets carry an MPDU payload for the MAC layer.

Figure 4–12:   Message sequence diagram for the energy detection mechanism

Figure 4–13 shows the structure of the PPDU.

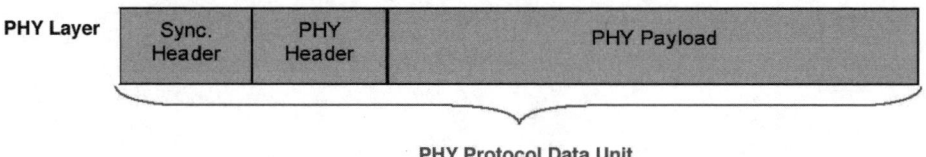

Figure 4–13: PPDU structure

Chapter 5 Medium Access Control Sublayer
... arbitrating channel access and more

The MAC sublayer, together with the Logical Link Control (LLC) sublayer, comprises the data link layer (also called layer 2) in the ISO/OSI model [26]. The MAC layer provides access control to a shared channel and reliable data delivery. In the case of WPANs, optimal use of the wireless media is desirable, because they operate in restricted unlicensed bands shared by several other standard and non-standard wireless technologies, including WLANs. IEEE Std 802.15.4 uses a carrier sense multiple access with collision avoidance (CSMA-CA) algorithm, which requires listening to the channel before transmitting to avoid collisions with other on-going transmissions (a kind of wireless etiquette!).

The IEEE Std 802.15.4 MAC sublayer is in charge of several other functions such as generation of acknowledgment frames, association, disassociation, security control, and optional star network topology functions such as beacon generation and guaranteed time slot management. The IEEE Std 802.15.4 MAC was designed to allow the implementation of a very thin (simple) protocol stack. This facilitates the quick development of applications and has a direct impact on improving power consumption — the secret is simplicity!

The set of standards from IEEE 802 differ at the physical layer and MAC sublayer, but share a common interface at the data link layer. This is achieved through the standardized IEEE 802.2™ LLC. This standard LLC interface was defined prior to the introduction of wireless networking standards, i.e., WLAN and WPAN. For this reason, IEEE Std 802.15.4 defines the Service Specific Convergence sublayer, which makes the appropriate interface between the MAC and IEEE Std 802.2, while allowing the definition of other LLCs more appropriate for wireless media.

The IEEE 802.15.4 MAC sublayer definition contains enhanced functionality normally located in the LLC, making it suitable to be interfaced with the network layer directly, allowing the simple implementation of wireless devices.

The IEEE Std 802.15.4 MAC provides application support for the creation of two wireless network topologies: star topology and peer-to-peer. The management of these types of networks is done in the network layer and is beyond the scope of IEEE Std 802.15.4. In this regard, the MAC only performs functions required by prospective network layers. Several application scenarios, mostly in the home networking segment, drove the vision for star networks, while industrial and commercial applications drove the vision for peer-to-peer networking. The peer-to-peer topology enables the creation of the more exotic ad hoc, self-organizing wireless networks (principally multihop/ mesh configurations).

All devices, regardless of topology, participate in the network using their unique IEEE 64-bit addresses; this address can be exchanged for a short 16-bit address allocated by the PAN coordinator thus ensuring that it is unique within its network; this process is managed by the association procedures explained below.

Star topology

In the star topology, communication is controlled by a unique PAN coordinator that operates as a network master, sending beacons for device synchronization (including superframe control) and maintaining association management. In this topology, the network devices communicate only with the PAN coordinator, as shown in Figure 5–1.

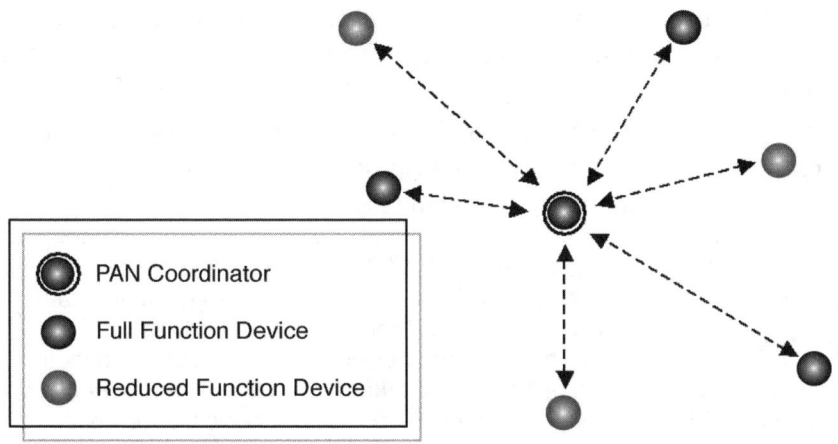

Figure 5–1:   Star network topology

Any full-function device (FFD) may establish its own network by becoming a PAN coordinator. Star networks operate independently from their neighboring networks. The PAN coordinator of a new star network in formation must choose a network identifier that is not used by any other network in its vicinity. After this procedure is performed, the PAN coordinator can start sending beacons and permitting association to network devices requesting it.

Network devices wishing to join a star network must listen for beacon messages sent by the PAN coordinator. After a beacon is received, the network device can send an association request back to the PAN coordinator, which in turn makes the decision to allow the association or not.

Star networks also support a non–beacon-enabled mode. In this case, the PAN coordinator uses beacons for association purposes only. Network device synchronization is achieved by polling the PAN coordinator for data on a periodic basis.

Peer-to-peer topology

The peer-to-peer topology allows any FFD to communicate with any other FFD within its range, and to have messages relayed to FFDs outside its range, via multi-hop routing of messages. This topology enables the formation of more complex networks, including ad hoc, self-organizing, and self-healing structures. IEEE Std 802.15.4 does not specify the details of any of these networks; it only defines the MAC functionality to enable this capacity.

RFDs may be employed in a peer-to peer network, but only as peripheral devices, since they lack the capability to relay packets. A number of FFDs sufficient to connect the network must be present. Figure 5–2 shows a typical peer-to-peer network.

Peer-to-peer communications requires additional device memory due to the size of the routing tables employed in the network layer.

In one type of peer-to-peer topology, the cluster-tree network, the network can have a variable number of network coordinators that can serve as cluster managers or "cluster heads" within a cluster-tree structure. The main advantages of a multi-clustered wireless network are the increased area coverage possible with multi-hop communication and improved reliability due to the inherent redundancy of devices that can relay messages through several paths. Figure 5–3 illustrates an example of a typical cluster-tree network; the figure shows network devices, cluster-heads, and the PAN coordinator that operates as the director of the operation for the entire network.

On network formation, each device in a peer-to peer structure needs to proactively search for other network devices. Once a device is found, the two devices can exchange parameters to recognize the type of services and features each support.

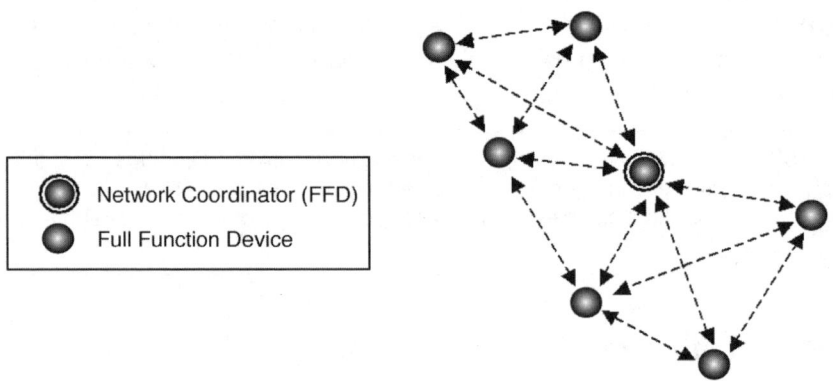

Figure 5–2:   Peer-to-peer network topology

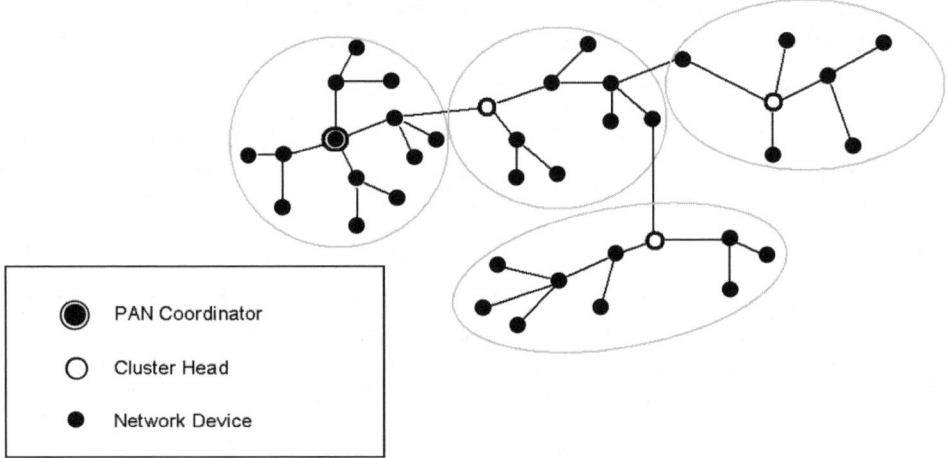

Figure 5–3:   Cluster-tree network topology

SUPERFRAME STRUCTURE

IEEE Std 802.15.4 allows the implementation of an optional superframe structure. The superframe is managed by the PAN coordinator and is bounded by beacon messages sent by the PAN coordinator at programmable regular intervals. Each beacon contains information that will help network devices synchronize to the network; this information includes the network identifier, beacon periodicity, and superframe structure. A superframe is divided in 16 contiguous time slots; the first time slot starts at the end of the beacon frame.

Network devices that wish to communicate with the PAN coordinator must attempt to do it in the time between two successive beacons. This period of time is called the contention access period (CAP). To communicate with the PAN coordinator, each network device needs to access the channel using CSMA-CA. Figure 5–4 shows the generic structure of a superframe.

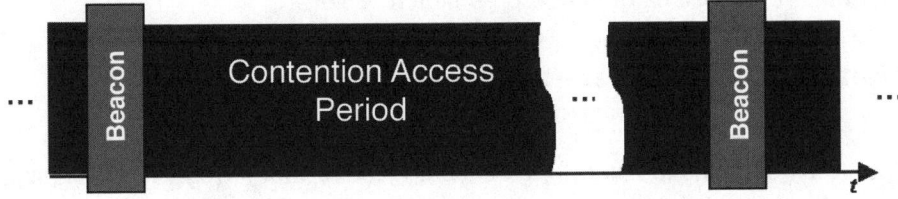

Figure 5–4: Generic superframe structure

The PAN coordinator can assign dedicated portions of the superframe to a specific network device requesting it. These segments of time are called guaranteed time slots (GTSs). This capability supports applications with a particular bandwidth requirement or needing lower communications latency. GTSs are all grouped toward the end of the superframe as is illustrated in Figure 5–5. The length of time covering all GTSs is defined as the contention-free period (CFP).

Each GTS is formed by an integer multiple of time slots. Each time slot is equal to 1/16 of the time between the start of two successive beacons.

Figure 5–5: Superframe structure with GTS

MAC DATA TRANSFER MODEL

The nature of the radio-frequency spectrum offers several challenges to the wireless communication system designer. The IEEE Std 802.15.4 MAC was designed to take these challenges into account and provide measures to improve communications reliability by offering a fully acknowledged protocol using CSMA-CA and a frame integrity check.

The data transfer model of IEEE Std 802.15.4 depends on the network topology in use. In star networks the communication exchange always occurs between a PAN coordinator and a network device while in the peer-to-peer mode, a device communicates with any others in its vicinity.

Star networks can have two types of data transfer mechanisms depending on whether the PAN coordinator is beacon-enabled or not.

Star Network Data Transfer—Beacon-enabled

In a beacon-enabled star network, a network device wishing to send data to the PAN coordinator needs to listen for a beacon. If it does not have a GTS assigned, the device transmits its data frame in the contention access period, in accordance with the CSMA-CA procedure. If the device has a GTS assigned, it waits for the appropriate point within the superframe structure to transmit its data frame. After receiving the data frame, the PAN coordinator sends back an acknowledgment to the network device; at which point the data transfer is finished. The message sequence diagram for this process is shown in Figure 5–6.

Figure 5–6: Star network data transfer from network device — beacon-enabled network

When the PAN coordinator has data pending to be transmitted to a network device, it sets a special flag in its beacon. Once the appropriate network device detects that the PAN coordinator has data pending for it, it sends back a data request message. The PAN coordinator responds with an acknowledgment followed by a data frame. The transaction is finished by an acknowledgment sent from the network device. The message sequence diagram of this process is summarized in Figure 5–7.

Then IEEE Std 802.15.4 design team decided not to implement a combined acknowledge/data frame in order to simplify implementation.

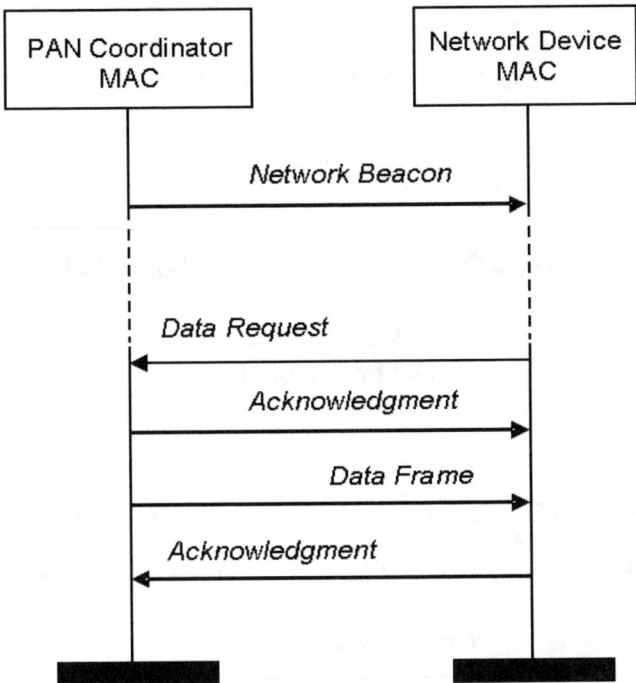

Figure 5–7: Star network data transfer from PAN coordinator — beacon-enabled network

Star Network Data Transfer—Non-Beacon-enabled

In a non-beacon-enabled star network, a network device wishing to transfer data sends a data frame to the PAN coordinator using CSMA-CA. The PAN coordinator responds to the network device, sending an acknowledgment message, as shown in Figure 5–8.

Similarly, when the PAN coordinator requires making a data transfer to a network device, it shall keep the data until the network device sends a data request message. The acknowledge message from the PAN coordinator will contain information indicating the network device if there are data pending; in which case, the data will be sent immediately after the acknowledgment. The transaction is finalized when the network device acknowledges reception of the data frame, as indicated in Figure 5–9.

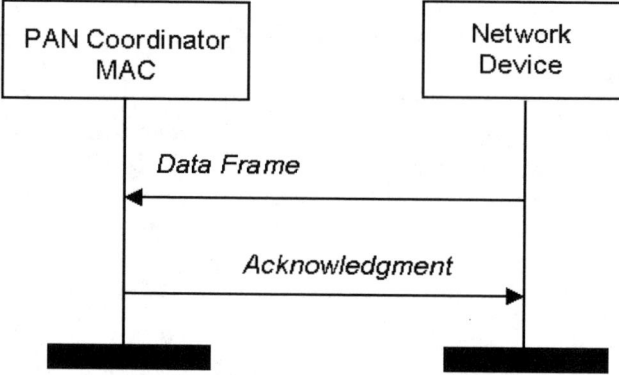

Figure 5–8: Star network data transfer from network device — non-beacon-enabled network

Figure 5–9: Star network data transfer from PAN coordinator — non-beacon-enabled network

Peer-to-Peer Data Transfer

For peer-to-peer topologies the data transfer strategy is governed by the specific network layer managing the wireless network. A given network device may stay in reception mode scanning the radio-frequency channel for on-going communications or can send periodic "hello" messages to achieve synchronization with other potential listening devices.

MAC Services

The MAC sublayer provides two services to the higher layers. These services are the MAC data service and the MAC management service [(called the MAC sublayer Management Entity or MLME)] and are accessed by the MAC common part sublayer service access point (MCPS-SAP), and the MAC management service access point (MLME-SAP), respectively. For each of these services, IEEE Std 802.15.4 defines a set of protocol primitives that enables the full functionality of LR-WPAN devices. Figure 5–10 shows the detailed IEEE Std 802.15.4 protocol stack architecture and the virtual link between MAC peer entities.

The services of a layer are the capabilities it offers to the next higher layer (or sublayer). The upper layers build their functionality based on the services the lower layers present to it.

The MAC services allow the transport of protocol units between peer entities.

Transmission Scenarios

The packet exchange between two transceivers is susceptible to errors due to the nature of the wireless medium. When one or more errors are present in any given packet, the message will not arrive at the intended MAC recipient either because the recipient PHY layer rejected the packet or the PHY could not synchronize (correlate) with the message itself (in this case, the PHY did not "hear" the message).

In any acknowledged data transfer there are three possible transmission scenarios, as follows:

- *Successful Data Transmission:* The originator device sends a message to a recipient MAC. The message is received with no errors detected and an acknowledgment message is sent back to the originator before a timeout occurs. A successful data transmission scenario is illustrated in Figure 5–11.
- *Lost Message Frame:* A message addressed to an intended MAC recipient never arrived at its destination. In this case, the originating device will declare timeout and will make another attempt to send the message. After a number of attempts, the MAC of the originating device will send a notification to its upper layer indicating transmission failure. This transmission scenario is shown in Figure 5–12.

- *Lost Acknowledgment Frame:* An acknowledgment frame was not received by the originator of the message transaction. Similar to the previous case, after a number of attempts, the MAC of the originating device will send a notification to its upper layer indicating transmission failure. This transmission scenario is shown in Figure 5–13.

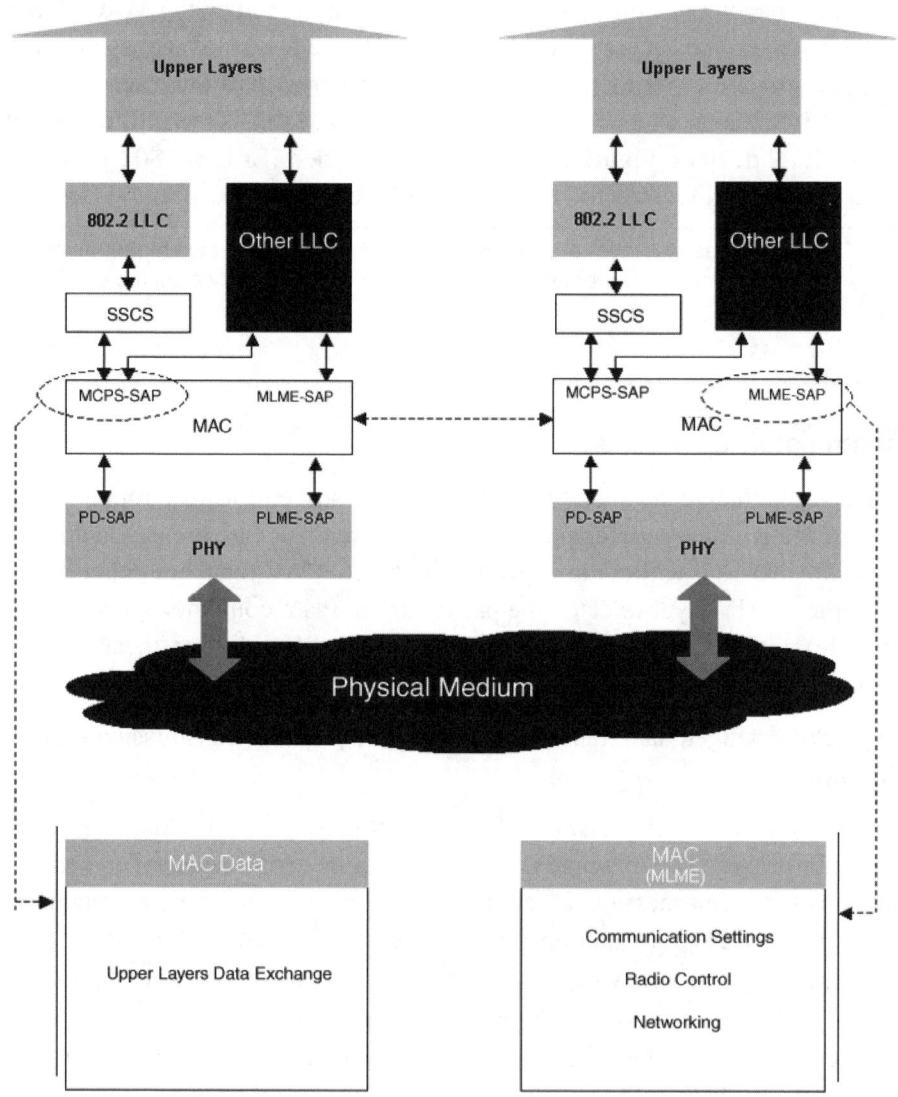

Figure 5–10:   IEEE Std 802.15.4 protocol stack architecture

Figure 5–11: Transmission scenario — successful data transfer

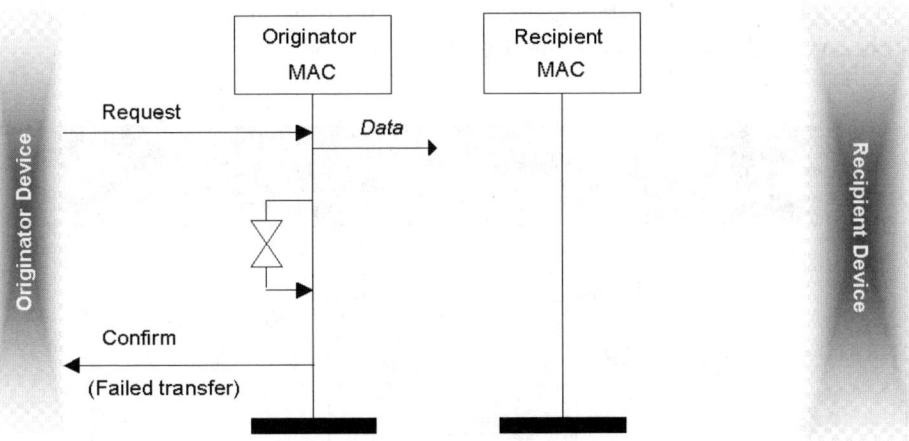

Figure 5–12: Transmission scenario — lost message frame

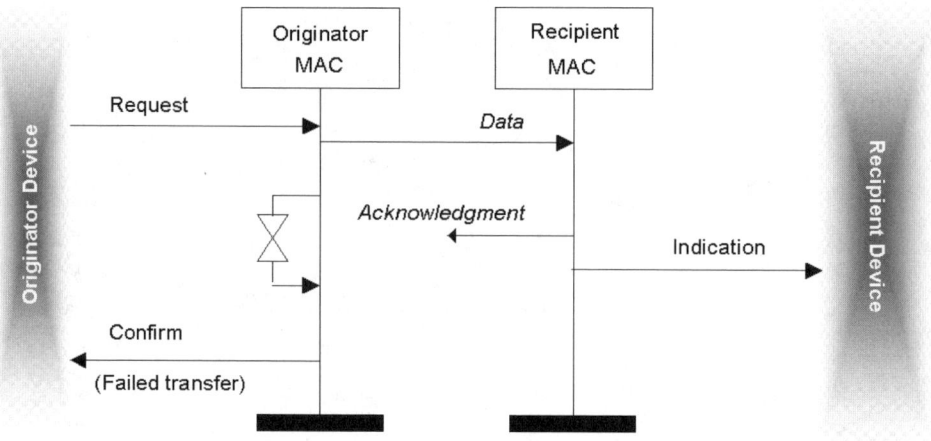

Figure 5–13: Transmission scenario — lost acknowledgment frame

MAC Data Service

The MAC data service provides three primitives for data transfer: data.request, confirm, and indication as shown in Figure 5–14. This figure shows the sequence diagram for a typical data transfer between two devices. The asynchronous design of this service does not require the use of a data.response primitive.

Figure 5–14: Message sequence diagram for data exchange mechanism

The interface offered by the MAC data service has the capability of diverse addressing schemes that enable simple implementation of star and peer-to-peer network topologies. In this sense, the MAC enables the upper layers to control which address fields are present in the packet to be transmitted, allowing the implementation of diverse type of network layers, including the most exotic multi-hop ad hoc networks. IEEE Std 802.15.4 uses a standard IEEE 64-bit address and short allocated addresses defined by association mechanisms. In addition, the MAC data service supports two optional primitives that enable the upper layers to purge an MSDU from the MAC queue. These primitives are MCPS-PURGE.request and MCPS-PURGE.confirm.

MAC Management Service

The MAC management service provides support for commands to control communication settings, the radio control, and networking functionality. The MLME primitives are summarized in Figure 5–15. The following paragraphs provide an overview of these primitives.

Primitive	Category	Description	Request	Confirm	Response	Indication
GET	Communication Settings	MAC PAN information base management	X	X		
SET			X	X		
RESET			X	X		
RX-ENABLE	Radio Control	Enables/Disables radio system	X	X		
SCAN		Scan radio channels	X	X		
ASSOCIATE	Networking	Association control with a network coordinator	X	X	X	X
DISASSOCIATE			X	X		X
GTS		GTS Management	X	X		X
ORPHAN		Orphan device management			X	X
SYNC		Control of device syncronization with network coordinator	X			
SYNC-LOSS						X
START		Beacon Management	X	X		
BEACON-NOTIFY						X
POLL		Beaconless Syncronization	X	X		
COMM-STATUS		Communication Status				X

Figure 5–15: MAC management service primitives

To enable the implementation of very low complexity devices, some of the MAC management primitives are optional (i.e., they do not need to be implemented in order for a product to be IEEE Std 802.15.4 compliant). These primitives are MLME-GTS, MLME-ORPHAN, and MLME-STAR. As will be shown later, these primitives provide support for the star network topology.

MAC PAN INFORMATION BASE MANAGEMENT PRIMITIVES

The MAC PAN Information base (PIB) contains configurable attributes to manage the MAC sublayer. These attributes can be read by the MLME-GET primitive and written by the MLME-SET primitive. In addition, the attributes can be reset to their default values using the MLME-Reset primitive.

The MLME-Reset primitive also resets the MAC sublayer.

Figure 5–16 shows the message sequence diagrams for the procedure of reading or writing the PIB attributes.

Figure 5–16: Message sequence diagram for MAC PIB reading and writing mechanism

ENABLING AND DISABLING THE RECEIVER

The radio receiver can be enabled or disabled by means of the MLME-TRX-STATE primitive. The action can be executed immediately or scheduled for a later time. The purpose of scheduling is to enable the MAC to offer network synchronization capabilities to the upper layers.

This primitive allows the power consumption to be lowered in systems that use a network layer that takes advantage of this feature i.e., scheduling data exchange at determined instants, and minimizing power consumption. Figure 5–17 shows the message sequence diagram for this primitive.

Figure 5–17: Message sequence diagram for transceiver enabling mechanism

SCANNING RADIO CHANNELS

MLME-SCAN allows the initiation of a radio channel scan over a list of available channels. There are four types of channel scans. They are as follows:

- *Energy Detection Scan:* This scan allows measuring the radio-frequency energy in each of the logical channels specified. The measurement is performed by the MLME by issuing a PLME-ED.request primitive.

  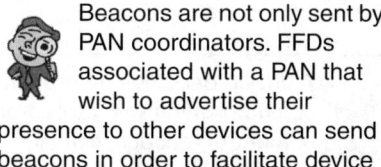 Beacons are not only sent by PAN coordinators. FFDs associated with a PAN that wish to advertise their presence to other devices can send beacons in order to facilitate device discovery in non-star networks.

- *Active Channel Scan:* This scan searches for beacons being transmitted in the radio sphere of influence of the device. For each logical channel, the device first sends a *Beacon Request* Command (see MAC Command Frames on page 87), which causes any prospective PAN coordinator (i.e., an FFD starting the network formation process) to send a beacon.

- *Passive Channel Scan:* Search for beacons being transmitted in the radio sphere of influence of the device without sending a *Beacon Request* Command.

- *Orphan Channel Scan:* Allows an orphan device (a device that has lost connection with its coordinator) to perform a scan to locate the beacons of its PAN. This search is performed over the specified list of logical channels.

Each of the above type of channel scans operates following the message sequence diagram shown in Figure 5–18.

Figure 5–18: Message sequence diagram for the channel scanning mechanism

A device becomes an orphan when it determines that it has lost communication with its coordinator. Communication may be lost for a number of reasons including:

- Fading or interference on the existing channel
- A switch of the coordinator to another channel, due to degradation of the existing channel (e.g., interference)
- Motion of either the device or the coordinator, so that the two are now out of range of each other.

ASSOCIATION AND DISASSOCIATION CONTROL

Following a successful active or passive channel scan, a network device can issue an association request (MLME.ASSOCIATE.request) to the coordinator found.

Depending on the capabilities and the requirements of the application controlling the PAN coordinator, this one can accept or reject the association request by issuing an MLME-ASSOCIATE.response with the appropriate parameters.

The MAC of the device requesting association sends an *Association Request* command that is acknowledged by the PAN coordinator. The acknowledgment of this message does not imply that the association was accepted; it simply implies a confirmation that the command was received. After receiving the association request, the MAC of the PAN coordinator needs to determine if it has sufficient resources to allow another device in its network. The PAN coordinator needs to make this decision in the time specified in the *aMaxAssociationPeriod* (a MAC constant parameter) or the requesting device will declare a timeout.

The details of the process for selecting a given PAN coordinator after a channel scan are not defined in the IEEE 802.15.4 standard. The upper layers of a particular implementation must define an association strategy.

Another important feature of the association process is the ability to request a short 16-bit allocated address to the PAN coordinator. This enables better bandwidth utilization because it reduces the total length of the packet (16-bit as opposed to 64-bit addresses). If the associating device does not request a 16-bit address allocation the device will participate in the network using its extended 64-bit unique address. Figure 5–19 illustrates the message sequence diagram of the association process.

The disassociation process can be initiated by the network device or the PAN coordinator by using the MLME-DISASSOCIATE primitives. The message sequence diagram of this process is shown in Figure 5–20.

GUARANTEED TIME SLOT MANAGEMENT

As described previously, IEEE Std 802.15.4 supports the optional use of a superframe, which enables the implementation of guaranteed time slots. The MLME-GTS primitives allow the allocation of new GTS, deallocation of an existing GTS, or reallocation of a GTS to eliminate time slot fragmentation.

A single GTS can extend over one or more superframe time slots. The management of GTSs is performed only by the PAN coordinator, which controls how many of the 16 available time slots are assigned to the contention-free period (the rest are assigned to the contention access period). The PAN coordinator may allocate up to seven GTSs.

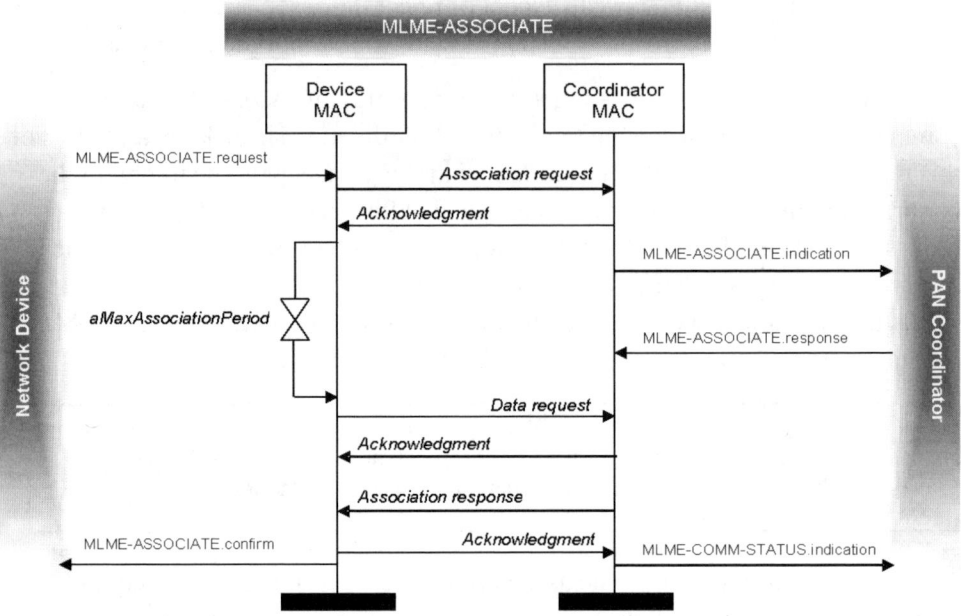

Figure 5–19: Message sequence diagram for association mechanism

Figure 5–20: Message sequence diagram for the disassociation mechanism

> The optional GTS capability allows the implementation of wireless links between a PAN coordinator and a network device with a guaranteed throughput (assuming a reliable link).
>
> The PAN coordinator can potentially assign a single GTS covering the whole contention-free period, which in turn may occupy the whole superframe period, less the minimum length contention access period of 440 symbols.

The GTS allocation/deallocation mechanism can be initiated by the network device or by the PAN coordinator as indicated in the GTS message sequence diagram of Figure 5–21. On reception of a *GTS request* message, the PAN coordinator determines if space is available within the superframe structure based on the remaining length of the contention period and the number of time slots requested.

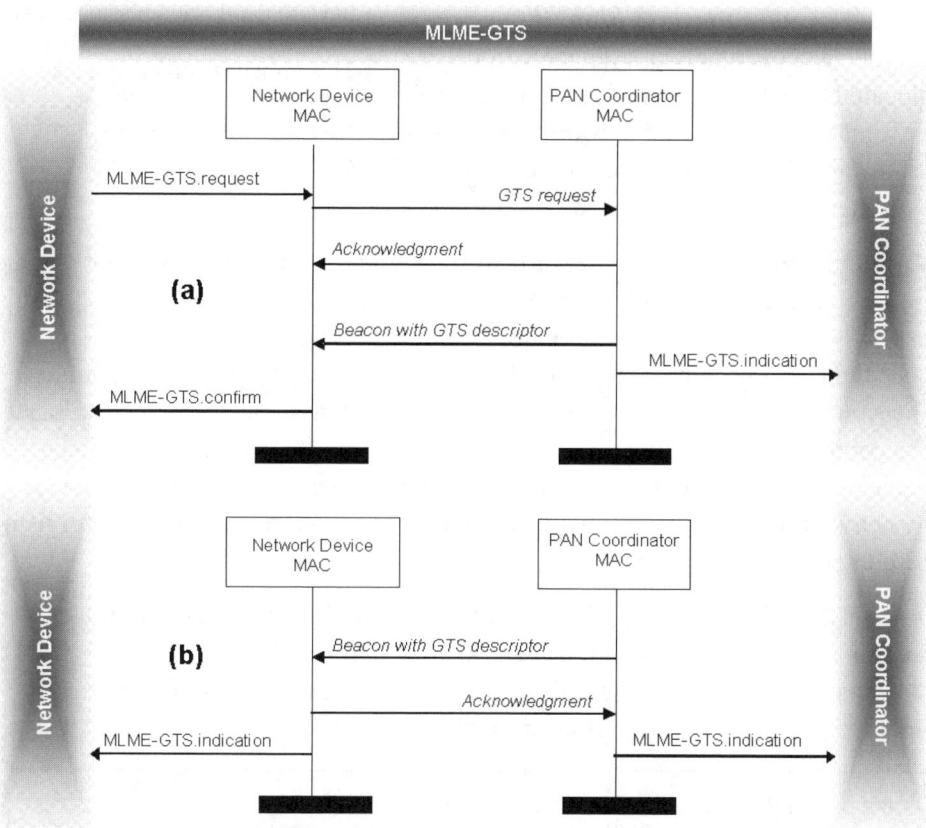

Figure 5–21: GTS allocation/deallocation procedure (a) initiated by the network device and (b) initiated by the PAN coordinator

When the PAN coordinator is responding to a GTS request, it will generate a beacon with its GTS fields indicating the time slot allocated and the number of slots assigned; if there is no capacity in the superframe to allocate a new GTS of the desired characteristics, the PAN coordinator will deny the request. In this case, the GTS fields in the beacon will indicate the number of time slots remaining in the contention-free period.

 Transmissions from network devices in a GTS do not require the use of CSMA-CA.

The GTS message exchange mechanism is fully acknowledged. It concludes with the generation of an MLME-GTS.confirm primitive that includes all relevant information of the GTS allocation procedure. After being assigned a GTS, a network device will use its allocated time to communicate with the PAN coordinator, although the network device will also be able to communicate with the PAN coordinator using the contention access period.

 When a network device generates a data request in a GTS-enabled network, the device will have to defer transmissions until the start of the assigned GTS. This transmission cannot take more time than the allocated time (number of allocated time slots).

A network device or the PAN coordinator can request the deallocation of an existing GTS. If the operation is initiated by the network device, its MLME should be instructed via an MLME-GTS.request with the appropriate parameters. Similarly, when the PAN coordinator initiates the deallocation of a given GTS, it should send a beacon with its GTS fields indicating that the GTS assigned is being deallocated. The complete deallocation process follows the same steps as the allocation mechanism as presented in Figure 5–21. After the deallocation of a given GTS, the network device can continue communicating with the PAN coordinator using the contention access period.

On a GTS-enabled network, the contention-free period of the superframe can become fragmented as a result of several allocations and deallocations. The PAN coordinator is in charge of removing gaps within the contention-free period in order to ensure optimal use of bandwidth.

If a device loses synchronization with the PAN coordinator, the GTS will be deallocated. After reestablishing a connection with the PAN coordinator (following a scan process), the device may request GTS allocation again.

ORPHAN DEVICE MANAGEMENT

In the event that a network device loses contact with its PAN coordinator, it will perform an orphan channel scan using the MLME-SCAN primitive. The MAC of the network device will react by sending *orphan notification command* messages through its list of available channels. When the MAC of the PAN coordinator receives this notification, it generates an MLME-ORPHAN.indication, which causes the PAN coordinator to verify if the network device was previously associated with its network. If the network device was associated to the PAN coordinator, it will generate an MLME-ORPHAN.response. At this point the MAC of the PAN coordinator will send a *coordinator realignment* command. In the event that the network device was not previously associated with the PAN coordinator, there will be no response from it. Figure 5–22 shows a message sequence diagram for the orphan notification mechanism.

A network device does not need to implement the MLME-ORPHAN primitives.

Figure 5–22: Message sequence diagram for orphan notification mechanism

SYNCHRONIZATION CONTROL

Synchronization control with a PAN coordinator is achieved with the use of the MLME-SYNC and MLME-SYNC-LOSS primitives. MLME-SYNC allows the network device to search for beacons in a beacon-enabled PAN or to poll for the PAN coordinator for any pending data in a non-beacon-enabled PAN. The search process is initiated with an MLME-SYNC.request. The search is done by activating the radio receiver and waiting, for a given amount of time, for a beacon frame to arrive from the PAN coordinator.

In a beacon-enabled PAN, the search for beacons can be performed in one of two modes: a mode in which the MAC of the network device will continuously track the PAN coordinator beacons; or a mode in which the MAC will locate the beacon only once. In both cases, if a beacon is received indicating data pending for the receiving network device, it will send a data request command to the PAN coordinator. In a non-beacon-enabled PAN, the MLME-SYNC.request will cause the MAC of the network device to generate a data request command to poll the PAN coordinator for any pending data. The message sequence chart for network device synchronization is shown in Figure 5–23.

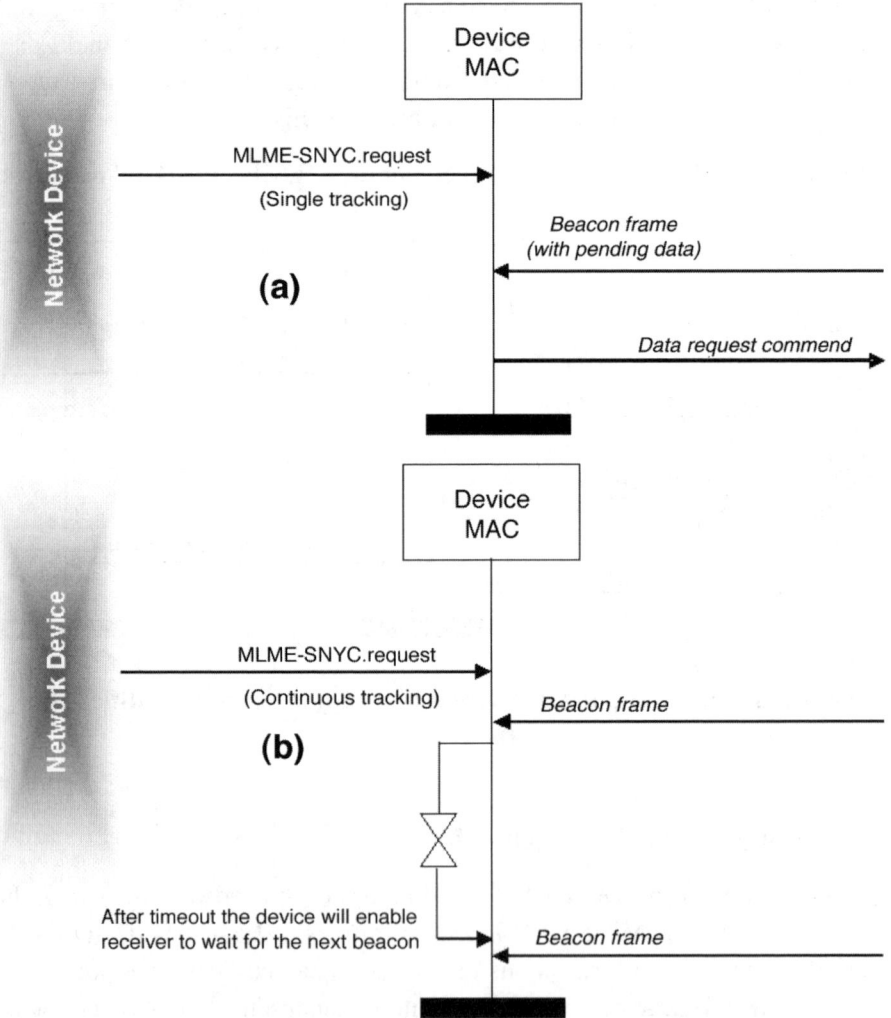

Figure 5–23: Message sequence diagram for network device synchronization (a) single tracking and (b) continuous tracking

In the event of a loss of synchronization with the coordinator, the MAC will generate an MLME-SYNC-LOSS.indication. Four possible situations can cause a loss of synchronization event. They are as follows:

- *Beacon Lost:* After MLME-SYNC, either initially or during tracking, the beacon message has not been received. The message sequence chart for this scenario is shown in Figure 5–24.
- *Coordinator Lost:* Several attempts to communicate with the coordinator have failed.
- *PAN ID Conflict:* The network device detected a PAN identifier conflict, defined as two different PAN coordinators with the same ID being in range of the network device.
- *Realignment:* The network device received a coordinator realignment command message from the PAN coordinator.

Figure 5–24: Message sequence chart for loss of synchronization due to beacon lost scenario

BEACON MANAGEMENT

Beacon generation is initiated with the use of the MLME-START primitive. The parameters of this primitive allow configuring the device as a PAN coordinator, selecting a logical channel, establishing beacon periodicity, and setting up the superframe characteristics. After the MLME-START.request, the MAC will respond with an MLME-START.confirm as illustrated in Figure 5–25.

Figure 5–25: Message sequence chart for procedure to start generating beacon transmissions

When a network device receives a beacon frame containing the PAN identifier of the network with which it is associated, the MAC will interpret its content. If the beacon frame contains one or more octets of payload (data), the MAC will issue an MLME-BEACON-NOTIFY.indication event. Figure 5–26 illustrates the beacon notification mechanism.

 Full-function devices operating in a peer-to-peer network can generate beacon messages to neighboring devices. These beacons can act as "Hello" messages during network formation. MLME-START with the parameter *PANCoordinator* set to FALSE enables this functionality. In this context, the device generating beacons is called simply a Coordinator.

In IEEE Std 802.15.4, only full-function devices have the capability of generating beacons.

Figure 5–26: Message sequence chart for beacon notification mechanism

> If the network device received a beacon generated by its PAN coordinator, the MAC will check its device address in the address list field from the beacon frame. If its address is listed, the MAC will generate a *data request* message to the PAN coordinator, as indicated in the star network data transfer (beacon-enabled).

BEACONLESS SYNCHRONIZATION

For non-beacon-enabled networks, a network device can use the MLME-POLL primitive to poll the PAN coordinator for pending data at the discretion of the network device. After a MLME-POLL.request, the MAC will generate an MLME-POLL.confirm with the results of the polling procedure. Figure 5–27 shows the message sequence chart for this primitive.

COMMUNICATION STATUS

With this primitive, the MAC generates a variety of communication status messages to its upper layer. The MLME-COMM-STATUS.indication is invoked when any of the following conditions occur:

- After a transmission caused by a "response" primitive.
- After reception of a frame that did not pass the security procedure.

Figure 5–27: Message sequence chart for polling the coordinator
(a) no data pending and (b) with data pending

MAC Frame Structure

The IEEE Std 802.15.4 MAC frame structure is designed in a way that reflects the protocol simplicity and flexibility while having the minimal elements to overcome the challenges of the wireless media. A MAC frame consists of three parts: header, variable length payload, and footer.

The MAC header contains a frame control field and an addressing field. The frame control field specifies the type of frame, security usage, and the format and content of the address fields. The frame control field also indicates if an acknowledgment from the recipient of the frame is required. The address field contains the source or destination addresses as specified in the frame control field.

> Depending on the type of MAC frame and the network topology in which a device is operating, a MAC frame can contain one or more of the following: the source (originator) PAN ID, source device address, destination PAN ID, and destination address. This feature enables the application to use diverse network topologies.
>
> In a star network, a network device only uses the originator PAN ID and originator device address.

The MAC payload contains information specific to the type of transaction being handled by the MAC and can be logically divided in several fields for use by upper protocol layers.

Finally, the MAC footer consists of a 16-bit frame check sequence (FCS) based on the standardized ITU-T 16-bit cyclic redundancy check (CRC) algorithm (formally called CCITT 16-bit CRC). The general MAC frame structure is shown in Figure 5–28.

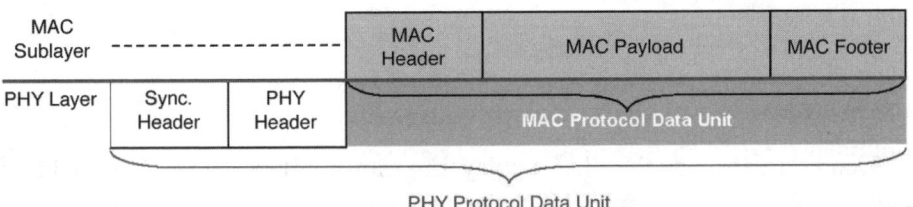

Figure 5–28: General MAC frame structure

When the three components of the MAC frame are assembled into the PHY packet, it is called the MAC protocol data unit (MPDU).

IEEE Std 802.15.4 defines four types of MAC frames: Beacon, Data, Acknowledgment, and MAC command. These frames are described in the following paragraphs.

Beacon Frames

In a beacon-enabled network, a full-function device may transmit beacon frames. In a beacon frame, the address field contains the source PAN ID and the source device address. The MAC payload of a beacon frame is divided into four fields:

- *Superframe Specification Field:* Contains the parameters that specify the superframe structure (if any).
- *Pending Address Specification Field:* Contains the number and type of addresses listed in the Address List Field.
- *Address List Field:* Contains a list of device addresses with data available at the PAN coordinator.
- *Beacon Payload Field:* Optional field that will contain broadcast data to the devices participating in its network within its range of coverage.

The format of the beacon frame is shown in Figure 5–29.

MAC Header			MAC Payload					MAC Footer
Frame Control	Sequence Number	Addressing Fields	Superframe Specification	GTS fields	Pending address field	Beacon Payload		FCS

Figure 5–29: Beacon frame format

Data Frames

Data frames are used by the MAC sublayer to transmit data. The address field will contain the PAN ID and device ID of the source and/or destination, as specified in the MCPS-DATA primitive. The format of the data frame is shown in Figure 5–30.

MAC Header			MAC Payload	MAC Footer
Frame Control	Sequence Number	Addressing Fields	Data Payload	FCS

Figure 5–30: Data frame format

Acknowledgment Frame

Acknowledgment frames are sent by the MAC sublayer to confirm successful frame reception to the originator of a message. The acknowledgment frame is generated only if the received message is requesting acknowledgment and the FCS (frame check sequence) is evaluated as good by the receiving device.

The acknowledgment frame does not contain the address field in the MAC header. Similarly, there is no MAC payload.

When a network device receives an acknowledge frame, it first verifies that it was expecting one, and then checks if the received frame sequence number matches the one it is expecting; otherwise, the acknowledgment frame should be discarded. Figure 5–31 shows the format of the acknowledgment frame.

> The acknowledgment was designed to be very short, in order to minimize network traffic. Recall that when a message acknowledgment is required, there will be one acknowledgment per MAC frame.

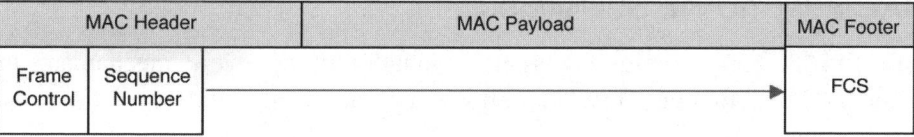

Figure 5–31: Acknowledgment frame format

MAC Command Frame

The MAC command frame is originated by the MAC sublayer and is in charge of all the MAC control transfers for each MAC command type shown in Table 5–1.

The MAC payload has two fields, the MAC command type and the MAC command payload. The MAC command payload contains information specific to the type of command in use. Figure 5–32 shows the details of the MAC command format.

MAC Header			MAC Payload		MAC Footer
Frame Control	Sequence Number	Addressing Fields	Command Type	MAC Command Payload	FCS

Figure 5–32: MAC command frame format

Table 5–1: MAC command frame types

Command Identifier	Command Type
1	Association Request
2	Association Response
3	Disassociation Notification
4	Data Request
5	PAN ID Conflict Notification
6	Orphan Notification
7	Beacon Request
8	Coordinator Realignment
9	GTS Request
10-255	Reserved

MAC FUNCTIONAL SCENARIOS

IEEE Std 802.15.4 contains a detailed explanation of the several mechanisms that define the MAC functionality. The following paragraphs provide a brief overview of these mechanisms.

Accessing the Channel

In IEEE Std 802.15.4, the physical radio channel is accessed using CSMA-CA before attempting to send any frame. The only exceptions are beacon frames, GTS transmissions, and acknowledgment frames sent in the superframe contention access period.

 In beacon-enabled PANs, slotted CSMA-CA is used, whereas in non-beacon-enabled PANs, the basic unslotted CSMA-CA algorithm is implemented.

Starting and Maintaining PANs

Network devices attempting to participate in a network must proceed to locate it by scanning the RF channels in its list of available channels. As previously described, the MLME-SCAN primitive allows performing active and passive scanning for beacons. After a suitable beacon is found, the network device initiates the association process.

Devices that lose communications with a PAN coordinator perform an orphan scan through its list of available channels. Once the PAN coordinator receives the inquiry from the orphaned device, it sends back a coordinator realignment command.

After a channel scan, if there are no appropriate networks to join, an FFD may begin operating as a PAN coordinator (sending beacons) through the use of the MLME-START primitive. In some instances, it is possible that two PAN coordinators with the same PAN identifier are collocated. One reason for this problem is that a PAN coordinator may self-assign its PAN identifier without consulting with neighboring coordinators.

A PAN coordinator detects a PAN identifier conflict if it receives a beacon with its own PAN identifier from a coordinator acting as a PAN coordinator or when a *PAN ID conflict notification* command is received from a device in its PAN. Similarly, a network device can detect a conflict when it receives a beacon with the same PAN identifier of the network it belongs to, in a time that has no correlation with the beacon period or when this beacon carries a source device address different from the address of its original PAN coordinator. In either case, the network device will send a *PAN ID conflict notification* command to its PAN coordinator.

After the PAN coordinator detects the conflict, it performs an active channel scan that enables the selection of a new PAN identifier and then it proceeds to broadcast a *coordinator realignment* command to its associated network devices.

 The procedure for a PAN coordinator to assign its PAN identifier is out of the scope of IEEE Std 802.15.4.

Device Synchronization

In a beacon-enabled network, devices are required to synchronize with the beacon frames in order to detect any pending messages. In a non-beacon-enabled network, devices may also send beacons acting as "hello" messages in order to enable discovery by neighboring devices.

GTS Management

The PAN coordinator is in charge of maintaining the complete administration of the superframe structure and the control of the allocation, deallocation, and reallocation of guaranteed time slots.

Network devices can request GTSs that extend over several superframe time slots. The PAN coordinator will allow the request depending on the availability of time slots in the contention-free period and the needs of the overall network. The MCPS-DATA.request primitive contains a parameter that indicates if the frame will be transmitted using a GTS or if it will be transmitted using the contention access period. When a GTS transmission is requested, the transmission is deferred until the start of the assigned GTS.

GTS deallocation can be initiated by a network device or the PAN coordinator. In both cases, a *GTS request* command is generated by the requesting device. GTS reallocation is only initiated by a PAN coordinator after detecting GTS fragmentation due to previous GTS deallocations.

SECURITY SERVICES

The MAC of IEEE Std 802.15.4 provides security services that are controlled by the MAC PIB. The MAC sublayer offers three security modes: an unsecured mode, an access control list (ACL) mode, and a secured mode.

A critical function of IEEE Std 802.15.4 MAC is frame security. Frame security is actually a set of optional services that may be provided by the MAC to the upper layers. The standard strikes a balance between the need for these services in many applications, and the desire to minimize the burden of their implementation on those applications that do not need them. The available services are as follows.

 Due to the variety of applications targeted by IEEE Std 802.15.4, the processes of authentication and key exchange are not defined in the standard.

Access Control

This security service enables the MAC to select the devices with which it is willing to communicate. The device may decide to communicate with some devices, and not others. To minimize complexity, the access control service is performed on an individual device basis, rather than on groups or collections of devices.

Data Encryption

Data encryption uses a symmetric cipher, in which the same key is used to encrypt plaintext at the message source and then decrypt the resulting ciphertext at the message destination. Devices without the key cannot decrypt the message. The standard defines encryption of beacon payloads, command payloads, and data payloads; other messages, such as acknowledgments, and message components, such as addresses, are not encrypted.

Frame Integrity

Frame integrity is a service that enables a receiving device to detect the modification of a message by parties without the correct cryptographic key, by appending a message integrity code (MIC) to the message. The standard defines integrity checking of beacon payloads, command payloads, and data payloads. Integrity is expected to be the security service most often used by IEEE Std 802.15.4 applications. Unlike file transfer or voice communication applications common with other protocols, IEEE Std 802.15.4 applications often transmit messages that do not convey secret information. What is often more important is that the messages are authenticated (i.e., the source is known and trusted), and that any tampering with or modification of the message is detected. For example, the fact that a light in a building is turned off is usually not secret; however, it is more important to be sure that any message to the light instructing it to turn off comes from a trusted source (e.g., the room switch), rather than a prankster or cracker.

Sequential Freshness

To avoid replay attacks to an IEEE Std 802.15.4 network, in which an old message is stored by an entity without the cryptographic key and then replayed later, an ordered sequence of values may be appended to the message. When received, this freshness value is compared against a stored value; if it is fresher than the stored value the freshness check passes and the new freshness value is stored. Although the freshness value can determine that one message is newer than another, it is a relative determination; no statement of absolute time is made. The use of the freshness service requires additional memory to store the present freshness value, and additional airtime to transmit the new freshness value. Although some applications may value the additional security provided by sequential freshness, other applications may find these requirements burdensome, so the use of freshness is optional.

IEEE Std 802.15.4 provides combinations of these four services in three security modes, which are designed to serve a wide range of applications.

Unsecured Mode

In unsecured mode, no security services are provided. This mode is suitable for some applications in which implementation cost is important, and security is either not required or obtained in other ways. Examples of this type of application include an advertising kiosk in a public place, or a low-power wireless motor controller in the center of a large, physically secure area outside of which its signal cannot be detected.

Access Control List (ACL) Mode

In ACL mode, the MAC maintains an access control list of device addresses with which it may communicate. The purported origins of received messages are compared against this list, and the result is forwarded to the upper layers along with the messages. The upper layers, in turn, may elect to ignore messages with purported origins not on the ACL. The ACL mode thus provides an authentication service. However, because the ACL mode offers no cryptographic security, it is possible for a message source address to be spoofed — that is, sent by a device with a differing address. If the ACL mode is used and spoofing is of significant concern in the application, additional cryptographic security in the upper layers should be employed to confirm the identity of the source device.

Secured Mode

In secured mode, the device may offer any of the four security services (access control, encryption, integrity, and freshness), depending on the security suite employed. Seven security suites have been defined as shown in Table 5–2.

Table 5–2: IEEE Std 802.15.4 security suites

Suite	Bits of integrity protection	Access Protection	Encryption	Integrity	Sequential Freshness
AES-CTR	0	Y	Y	N	Optional
AES-CCM-128	128	Y	Y	Y	Optional
AES-CCM-64	64	Y	Y	Y	Optional
AES-CCM-32	32	Y	Y	Y	Optional
AES-CBC-MAC-128	128	Y	N	Y	N
AES-CBC-MAC-64	64	Y	N	Y	N
AES-CBC-MAC-32	32	Y	N	Y	N

The AES-128 (Advanced Encryption Standard, with 128-bit keys and 128-bit block size) symmetric cryptography algorithm [24] is employed for all security suites.

The AES-CTR (AES Counter) mode employs the AES algorithm to provide access control and encryption services. The AES-CBC-MAC-128, AES-CBC MAC-64, and AES-CBC-MAC-32 (AES cipher block chaining message authentication code, with integrity codes of 128, 64, or 32 bits in length) modes employ AES to provide integrity and, optionally, sequential freshness (by the transmission of a freshness code). The AES-CCM-128, AES-CCM-64, and AES-CCM-32 (AES-CTR and CBC-MAC, with integrity codes of 128, 64, or 32 bits in length) modes employ AES to provide access control, encryption, integrity, and, optionally, sequential freshness (by the transmission of a freshness code). The length of the integrity code does not refer to the strength of the AES algorithm, but rather to the number of bits of the code actually transmitted. The optional sequential freshness code is five octets in length.

This set of security suites offers several advantages. A major advantage is that it employs only one cryptographic algorithm for all security suites. Other alternatives require additional algorithms, for example, a hash algorithm for integrity checking, to provide a complete security suite, resulting in a larger and, therefore, more costly, implementation if multiple security features are desired. In particular, by reusing AES in a clever way, the AES-CCM modes enable a single algorithm to provide all four security services in a very small implementation. In addition, the use of AES-CCM modes retains compatibility with other draft IEEE 802 standards, such as IEEE Draft Std 802.11i and IEEE Std 802.15.3. This not only allows circuit and code reuse between the standards; it ensures that a larger audience is reviewing and studying the security algorithms for weaknesses. Over time, this high level of interest increases confidence in the security algorithms.

If a device employs secure mode, at least AES-CCM-64 must be included in the implementation. This requirement was the result of a tradeoff between the desire for strong integrity protection and a desire to minimize the size increase of transmitted frames, especially if freshness is used. AES-CCM-128 has the strongest integrity protection, but it also adds 128 bits (16 octets), plus an optional 5 octets for the freshness code, to each transmitted frame, including beacons and other frames that may be relatively short, which could adversely affect battery life in some applications. Making AES-CCM-128 mandatory was therefore felt not to be suitable for the widest selection of applications. AES CCM-64, with only 8 octets added to each frame, was seen as a better compromise; a brute-force attack on a system sending AES-CCM-64 frames every 15.36 ms would still require an average of $4.5 * 10^9$ years to succeed.

Chapter 6 Network Functionality

...getting information from source to destination

IEEE Std 802.15.4 was designed to support multiple network topologies—from a star network to many network types based on peer-to-peer communication, including mesh, tree, cluster, and cluster-tree networks. Although the network layer *per se* is beyond the scope of the standard, this chapter describes some exemplary networks that may be constructed using IEEE Std 802.15.4, including samples of their associated message routing algorithms. Although message routing algorithms are beyond the scope of the standard (except for the simple MAC switching functionality of the star network), these algorithms are described in this chapter to illustrate the type of performance that may be achieved by IEEE Std 802.15.4 networks, and the wide variety of applications to which they may be applied.

FEATURE OVERVIEW

IEEE Std 802.15.4 supports star networks in a master–slave configuration and a very general form of peer-to-peer communication, which may be used to construct a variety of network types.

The star network (Figure 6–1) is a single- or double-hop network (i.e., all network devices are in range of a single common device), whereas peer-to-peer networks, for example, mesh (Figure 6–2) or cluster (Figure 6–3) networks, may be multi-hop networks (i.e., the message source and destination devices are not necessarily within range of each other, and communication may occur via multiple intermediate devices that relay messages). The standard thus supports both single-hop and multi-hop communication links.

Starting a Network

As described in Chapter 5, an IEEE Std 802.15.4 network is started by an FFD when it becomes the PAN coordinator. All networks must have exactly one PAN coordinator in each network. Figure 6–4 shows the network formation, which starts when an upper layer of the FFD sends the MLME-SCAN.request primitive to the MLME, requesting an active channel scan. Upon completion of the scan, the results are sent back to the upper layers via the MLME-SCAN.confirm primitive. If the results of the scan are acceptable, the upper layer selects a PAN identifier (which may be predetermined), and sends the MLME-START.request primitive (with the PAN coordinator parameter set to TRUE) to the MLME.

Figure 6–1:   Star network

Figure 6–2:   Mesh network

Figure 6–3: Cluster network

Figure 6–4: Network formation

The MLME-START.request primitive requires the MAC sublayer to place the PAN identifier *macPANId* value in the MAC PIB and the logical channel *phyCurrentChannel* value in the PHY PIB. After this is completed, the MAC sublayer sends the MLME-START.confirm primitive to the upper layers, the FFD begins operating as a PAN coordinator, and the network is started.

All IEEE Std 802.15.4 devices have a unique 64-bit IEEE address, *aExtendedAddress* (often called a "MAC" or "physical" address), stored in the MAC PIB table. When a device joins a network, it identifies itself by its *aExtendedAddress* value; at the discretion of the PAN coordinator, it may exchange that address for a shorter, network-specific logical address supplied by the PAN coordinator during the network association process.

UPPER LAYER NETWORK FORMATION POLICIES AND ALGORITHMS

As previously indicated in Chapter 5, some policies and algorithms associated with network formation and association are performed by upper layers of the protocol and not defined in the standard. A few alternatives for these algorithms are discussed here. Recall that what follows are only possibilities—other network-forming policies and algorithms could be used as well.

PAN Coordinator Selection

The first step in network formation is the selection of the PAN coordinator. The policy deciding the transmission of the MLME-START.request primitive (with the PAN coordinator parameter set to TRUE) from an upper layer to the MAC is, of course, determined by that upper layer. One may envision several application-dependent scenarios as follows:

- *Dedicated PAN coordinator:* There are some applications, for example, some home security systems, for which it is clear that only one device, the gateway device, needs to be connected to an outside network. This device should be the PAN coordinator. These applications may require the consumer to purchase the entire system as a whole at one time, so that the manufacturer may have complete control over the final network design and behavior. In other situations, the consumer may have the flexibility to purchase network devices individually. This policy may be reinforced by employing only one FFD in the network (as the PAN coordinator) and populating the rest of the network with RFDs. In such a network, exactly one device is eligible to become the PAN coordinator. This must be clearly understood by the consumer user, who may otherwise attempt to make a network with zero, two, or more devices each attempting to become PAN coordinator, leading to unsatisfactory results.

- *Event-determined PAN coordinator:* In other applications, it is desirable to employ a large number of identical devices, any one of which may become the PAN coordinator given an external stimulus (for example, a button press by a user). In this case, every device must be an FFD.

- *Self-determined PAN coordinator:* For these applications, the goal is to establish a network, but it is less important which device in the network is the PAN coordinator. An example of this type of application is a location-determining network. The purpose of this network is to determine the relative location of each network device, possibly by means of a distributed algorithm. In this type of application, there may be no external gateway, so any network device may serve as the PAN coordinator. One way of forming this type of network is to employ a form of power-on-network formation, in which the upper layer of each device instructs the MAC to begin an active channel scan (the needed step prior to network formation) upon device power-up. The first device to complete this scan (which should produce a negative result if the network is not yet formed) and send the result to the upper layer may be rewarded with the MLME-START.request primitive, and become the PAN coordinator. As always, all prospective PAN coordinators must be FFDs.

Selection of the PAN Identifier

Once the channel scan has been completed and the results returned to the higher layers, the results should be evaluated and action taken by the higher layers.

In some applications, it may seem desirable from a security standpoint to fix the desired PAN identifier *a priori*, so that new network devices will be limited to joining only one particular network (that must be already established). Although possible with IEEE Std 802.15.4 networks, this policy should be discouraged, due to the relatively small number of PAN identifiers available and the unpredictable behavior (from the user's perspective) that could result if two networks with identical PAN identifiers were co-located. (There is a procedure for PAN identifier conflict resolution in IEEE Std 802.15.4.) An alternative method to achieve this level of desired security is to require all network devices upon association to check the 64-bit extended address of the PAN coordinator, and compare it with the desired address. In this way, association with the correct network, regardless of PAN identifier, can still be guaranteed.

If a new PAN is to be created in the same region as a pre-existing PAN (a case in which a pre-existing or newly placed FFD becomes a dedicated PAN coordinator for the new PAN), the new PAN coordinator must posses a different PAN identifier than all PAN coordinators it is capable of hearing. However, if the goal is for an FFD to become a member of any network, and assume PAN coordinator status only if no PAN identifiers were received during the channel scan, a number of scenarios are possible.

If exactly one desired PAN identifier has been received during the channel scan, the device may choose to attempt to associate with the device transmitting that PAN identifier. If more than one device was heard transmitting a desired PAN identifier, some type of selection algorithm must be employed to determine to which device association should be attempted. This algorithm may consider other information transmitted in the received beacons, previous history (perhaps previous association attempts with a particular device have failed), or may employ a simple deterministic or stochastic (e.g., round-robin, first-in, or random selection) process.

The final scenario, of course, is the case in which no other PAN identifiers were heard and, for application reasons, the device is not permitted to become a PAN coordinator. In this scenario, the device may return to sleep for an application-dependent period, waking later to repeat the active channel scan.

The Use of Beacons

Beacon use is required in IEEE Std 802.15.4 networks for network discovery (i.e., when the network wishes to attract new network devices). Beacons also are useful for coordination and synchronization of an active network (e.g., for GTS control), and in applications for which it is desirable to minimize coordinator-to-network device message latency, because the network device can readily check the message pending field of a beacon to determine when messages are waiting for it.

However, beacons can be disadvantageous in some applications. For example, if no messages are expected from the coordinator to the network device, and only light traffic is expected from the network device to the coordinator (for example, in a wireless light switch), beacon transmission will waste the power of the coordinator, and reception of the beacons will waste the power of the network device. A better solution is to stop the beacon transmissions and have the network device asynchronously transmit its (rare) data frames as they are generated. This approach is especially desirable if the network is in a star configuration, and the PAN coordinator may be externally powered. In this case, the PAN coordinator may operate constantly in receive mode, while the (presumably battery-powered) network device need only expend significant energy when it transmits a state change to the PAN coordinator. As a result, the battery life of the network device may be greatly extended. This is true even if occasional coordinator-to-network device traffic is expected, requiring the network device to poll the coordinator for data.

THE STAR NETWORK

A star network is a good design choice for applications that need to cover a limited physical area, so that a single device (the master, which will be the PAN coordinator) may be in range of all other network devices (the slaves). IEEE Std 802.15.4 star networks must employ an FFD as the PAN coordinator; however, the other network devices may be either FFDs or RFDs. Because star network devices communicate only with the PAN coordinator, they may have a degenerate ACL consisting of a single entry (that of the PAN coordinator), resulting in potential implementation cost savings over peer-to-peer networks, which must store an ACL entry for every peer with which they communicate. An IEEE Std 802.15.4 RFD, operating as a network device in a star network, can be a good alternative for simple point-to-point applications that require an extremely low-cost implementation.

Because star networks by their nature are single- and double-hop networks, message latency can be lower than in multi-hop networks. If latency is a critical performance metric of the desired application, a GTS may be assigned by the PAN coordinator to reserve time for a particular network device in each superframe, avoiding the delay of contention-based channel access and guaranteeing a fixed amount of bandwidth to the network device. (The network device must be an FFD to employ GTS.) In this way, an IEEE Std 802.15.4 star network can provide maximum message latencies as low as 15.36 ms, suitable for use in PC peripheral applications such as wireless mice and joysticks. In the extreme case, it is possible to extend a single GTS to encompass all time between beacons; this enables a single network device to have the entire channel bandwidth, in excess of 115.2 kb/s for the 2.4 GHz band, for relatively high-bandwidth applications.

Conceptually, message routing in star networks is viewed differently than message routing in peer-to-peer networks. Because the PAN coordinator of a star network can hear all network devices and directly controls access to the shared channel, routing in star networks is viewed as occurring in the MAC layer, as packet switching, rather than in the network layer as part of a peer-to-peer message routing algorithm. With this idea in mind, message routing in star networks has been defined in IEEE Std 802.15.4. In fact, routing at this level can be shown to have some implementation advantages, because simple message relaying need not involve the higher layers of the protocol.

PEER-TO-PEER NETWORKS

The peer-to-peer communication capability of IEEE Std 802.15.4 devices allow the creation of many types of peer-to-peer networks, each with their own advantages and disadvantages. A sampling of network types possible with IEEE Std 802.15.4 FFDs is described below.

The Flat "Mesh" Network Topology

The simplest type of peer-to-peer ad hoc network to envision is perhaps the flat "mesh" network, a network composed of a number of identical network devices, not all of which may be in range of any one device. Messages may be relayed from source devices to destination devices via a large number of routing algorithms.

Although the network devices in the mesh may be identical, one must be "more identical than the others" to perform the PAN coordinator function, give the network its PAN identifier, and control the association of new devices to the network. However, the PAN coordinator need not take a role in message routing.

A true mesh network has regularly spaced devices forming a repeating grid pattern of communication links; however, the definition used here, allowing an ad hoc device distribution, has become common.

Routing

A principal problem to be overcome in a flat mesh network design is the problem of addressing. Because the network is logically flat (i.e., there is no hierarchy) and there is no other grouping or organization of network devices, the address of a network device does not provide clues to the route needed to get a message to it.

Nevertheless, many routing algorithms have been devised for such networks, a few of which are listed here as follows:

- *Flooding:* The simplest method to route messages is to send a copy of each message to all network devices. Although this is a correct algorithm (i.e., it will in fact deliver the message to the intended recipient), it is not very efficient, especially for the large networks IEEE Std 802.15.4 can support. Much power can be wasted sending the message to unintended recipients. This method is seldom used in practical networks of any size to transmit data traffic, although it is needed in nearly all networks for at least some control and status message traffic.
- *The Bellman–Ford algorithm:* The Bellman–Ford algorithm requires all network devices to maintain a routing table that contains routing costs (usually the number of hops, although more complex cost accounting, such as available power of network devices in the route, can be used) of the optimum route to all

other network devices, plus the address of the first device in that route. Devices maintain their tables by exchanging them with all devices within their range, and then comparing entries by destination. The route of a message is determined by the source device in advance ("source routing"), and placed *in toto* in the MSDU, as part of the message payload. Relaying network devices follow these routing instructions to deliver the message. Although an elegant solution to the routing problem, the Bellman–Ford algorithm suffers from poor dynamic behavior. If a communication link is broken, for example, its convergence behavior to the new set of optimal routes is not good. In addition, because the routing tables must have as many entries as there are network devices in the network, the costs to exchange them become prohibitive once the network becomes large.

- *The GRAd (Gradient Routing for Ad hoc networks) algorithm:* The GRAd algorithm [19] requires all network devices to maintain a cost table that lists the routing cost to each potential destination device (as opposed to the Bellman–Ford algorithm which lists all network devices). However, cost tables are not exchanged among network devices, and messages are not sent via a unicast transmission to particular network devices along a route determined by the message source. Instead, messages are broadcast by the source to all network devices in range, along with the cost value to the destination found in the source's cost table. Any neighbor hearing the broadcast message, and having a routing cost for the destination less than that sent by the source (and therefore presumably closer to the destination), waits for a random period and then rebroadcasts the message, listing its own routing cost. Acknowledgment is passive; the acknowledgment to the transmitting device is detection of the retransmission by a neighboring device. All others ignore the message, except to record the routing cost and destination sent by their neighbor, which they use to update their cost tables. In this way, the message slides down a "cost gradient" to its destination. The GRAd algorithm requires each network device that may be a message destination to flood the network upon network association with a special announcement message, so that it will appear in the cost table of other network devices. Alternatively, specialized route request messages can be broadcast by devices with messages for destination nodes for which they have no cost value stored.

- *Ant algorithms:* Because they also are ad hoc and self-organizing, biological analogies to the routing problem have also been explored. Of particular interest is the communication paradigm of ants. Ants communicate by placing a pheromone trail along the ground. If, for example, a wandering ant finds food, it returns to the nest, placing the "food" pheromone on the ground as it does so. Other ants detect this trail, arrive at the food, and return to the nest, laying down pheromone trails of their own that diffuse together into a wide trail. The pheromone evaporates quickly; an unused trail becomes undetectable to other ants in a short time. In this way, the ants quickly develop a single path to the food that is used by almost all ants.

Because this use of pheromones is a distributed communication system in which simple behavior of individual ants results in complex behavior of the larger network, it has clear appeal to the network routing problem, especially in the case where there is a single message sink, but possibly many message sources. One approach is to model the information sink (i.e., the PAN coordinator, the destination of all messages) as the nest, the network devices sourcing messages as the food, the collection of intermediate (relaying) network devices as the ground between food and nest, and the messages themselves as the ants. The PAN coordinator periodically transmits a special announcement message (containing a "number of hops taken" field), which is passed from network device to network device by unicast transmissions. At each network device, the device stores the source identity, and the identity of the neighbor from which it received the message, in a table, along with a time stamp. The network device then increments the "number of hops taken" field in the message and forwards it to a different neighbor. Network devices perform the pheromone "evaporation" function by removing old entries from their tables.

When a "real" message is generated in the network, destined for the PAN coordinator, if the source has not received the special announcement message from the PAN coordinator recently (so that an entry is not present for it in its table), it sends the message to a random neighbor. This process is repeated until the message is sent to a network device that has a table entry for the destination. That device then sends the message to the neighbor from which it received the special announcement message, as recorded in its table. The message is now "on the pheromone trail," and it will eventually arrive at its destination.

The ant communication paradigm has appeal for particular networks in which it can be guaranteed that there will be only a few network devices acting as message destinations, because each destination must transmit regular announcement messages. Examples of such networks include wireless sensor networks, which typically have many source devices (sensors) and few destination devices.

The Cluster Network Topology

The algorithms above notwithstanding, routing in a flat network has unavoidable limitations for many applications, and it would be desirable to have a network in which the routing algorithm was assisted in some way by the logical structure of the network, especially as the number of devices in the network becomes large. (Note that the *physical* structure of the network may still be ad hoc, however). This could eliminate, for example, much of the announcement message passing of the ant algorithm, and the exchange of very large routing tables inherent in the Bellman–Ford algorithm.

One approach to this problem suitable for many applications is the cluster network. In a cluster network, there is the concept of a "parent–child" relationship between network devices (Figure 6–5). The network forms, as do all IEEE Std 802.15.4 networks, with the PAN coordinator as the first device in the network. When a new device associates to the PAN coordinator (and therefore to the network), it becomes the child of the PAN coordinator, and the PAN coordinator becomes the parent of the new device. Should a second device come into range of the first network device (but, perhaps, out of range of the PAN coordinator), the second device may join the network as a child of the first device. Network devices may have many children (and grandchildren), but only one parent (Figure 6–6).

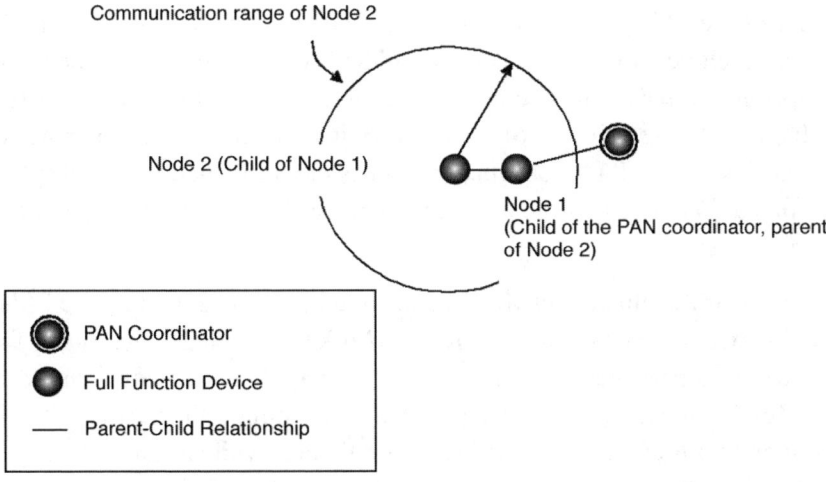

Figure 6–5:   Formation of a cluster network

Figure 6–6: Parents, children, and grandchildren in a cluster network

Discovery of the network by new devices is made possible by the transmission of beacons by each network device (including the PAN coordinator). When a new device appears, it may hear several beacons, and have several potential parents from which to choose. If (as is often the case) it is desirable to connect network devices as close to the PAN coordinator as possible, information may be placed in the beacon payload to assist the prospective network member in the parental selection process.

The structure of the cluster network is controlled to some extent by the PAN coordinator, which retains authority over network association (regardless of which network device a prospective member may contact). The PAN coordinator may prohibit, for example, a prospective member from joining the network at a device distant (many hops away) from the PAN coordinator, while allowing other prospective members to join closer to the PAN coordinator, in order to encourage a "flatter" cluster structure and control the latency of messages sent in the network.

One advantage of cluster networks is that it is straightforward to perform a periodic network status update, so that the PAN coordinator can be made aware of any broken communication links or missing network devices anywhere in the network. One way this can be accomplished is to have the PAN coordinator generate a "status update request" message, sent to all network devices that are not parents (i.e., all devices at the ends of the cluster branches). When received, every parent device relays this message to all of its children, so that all network devices either receive or relay this message. When a network device without children receives the status update request message, it replies with a "status update response" message, a very short message containing its network address sent to the PAN coordinator. Intermediate devices on the return path to the PAN coordinator "eavesdrop" on this message as they relay it, gaining information about network devices in their cluster, but perhaps out of range. They then append their information to the message (perhaps consolidating the messages received from several children) and forward it to their parent. This procedure is shown in Figure 6–7.

Figure 6–7: Network status update messages and routing in a network

A variant of this procedure is to require the status update response message to include the complete routing table of each network device. While lengthening the response message, this can make the PAN coordinator (and, in fact, any eavesdropping network device) aware of alternative routes to the same destination device, which may improve network reliability in the event of a network device or communication link failure.

Routing

Because their network devices have an understanding of network connections in their local area, routing in cluster networks can be more efficient than in flat networks. One efficient way to route in cluster networks is to use information gained in eavesdropping the status update response messages. By doing so, eavesdropping devices have a list of network devices "downstream" of the device from which the status response message was received. This information can be stored in a routing table.

A general routing algorithm for cluster networks is shown in Figure 6–8. The algorithm assumes that the network device stores a routing table, which consists of entries from two sources:

- Entries added by eavesdropping on devices within reception range. One of these entries will be for the parent of the device. These entries make up the "neighbor list," the list of all devices to which messages may be passed without intermediate relay.

- Entries added by eavesdropping on network status reply messages. These messages inform the device of all devices in the network "downstream," that is, away from the PAN coordinator.

Following the algorithm shown in Figure 6–8, a network device checks first if the destination device has an entry in its routing table. If an entry is found, the message is passed to the appropriate device. If an entry is not found, the message is sent to the parent, which, because it is closer to the PAN coordinator may have additional routing information by virtue of eavesdropping on network status response messages from a larger number of devices. In the extreme, a message may be relayed all the way to the PAN coordinator, but delivery is assured nevertheless.

A significant advantage of cluster networks for wireless sensor network applications is the small size of the network device routing tables, compared to those of flat networks. In general, for flat networks, a table entry is required for every potential destination; for a cluster network, the routing table may be much smaller because destinations not found in the table are still routed, via the parent.

A small table size reduces the memory requirement of the network devices, which can be a major factor in their product cost.

A disadvantage of cluster networks is the non-uniform distribution of message traffic among the network devices. Some devices, especially those logically close to the PAN coordinator, may have significantly higher traffic than devices further away from the PAN coordinator, leading to unequal battery life among network devices. Should the power supply of a device become exhausted, a network partition could occur. Several methods have been proposed to mitigate this effect, including the rotation of PAN coordinator duties among alternative network devices and the use of multiple routing table entries for the same destination.

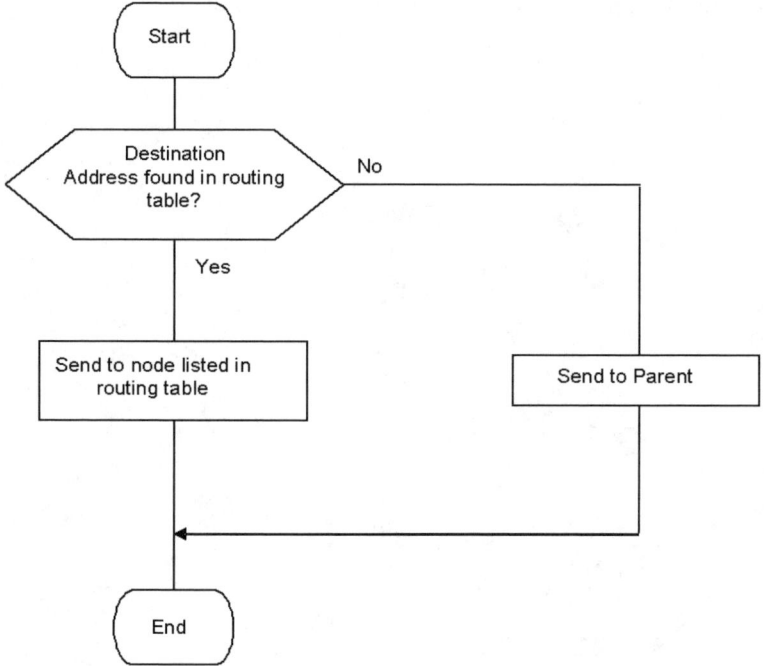

Figure 6–8: A message routing algorithm for cluster networks

The Cluster Tree Network Topology

The address space of IEEE Std 802.15.4 MAC is capable of supporting a large number of network devices. However, as the number of network devices continues to grow, even the routing tables of devices in a cluster network may grow to an impractical size, because a device's routing table must contain an entry for each of its offspring (child, grandchild, etc.).

To address this issue, hierarchy may be employed. A large network may be broken up into several, smaller clusters, connected in a hierarchical tree. An example of such a "cluster-tree" network is shown in Figure 6–9.

The large network shown in Figure 6–9 is composed of four smaller clusters, each with a "cluster head," or coordinator. The coordinator for Cluster 0 is, of course, the PAN coordinator. If the PAN coordinator has a set of rules governing network formation, the clusters may be formed as a natural part of the network formation process. These rules may take the form of a limit on the number of devices in a cluster, a limit on the number of hops any device may be from a coordinator, or a more sophisticated algorithm. Should the attempted association of a candidate device to a network device violate these rules, the PAN coordinator may allow the device to join, but only as the coordinator of a new cluster.

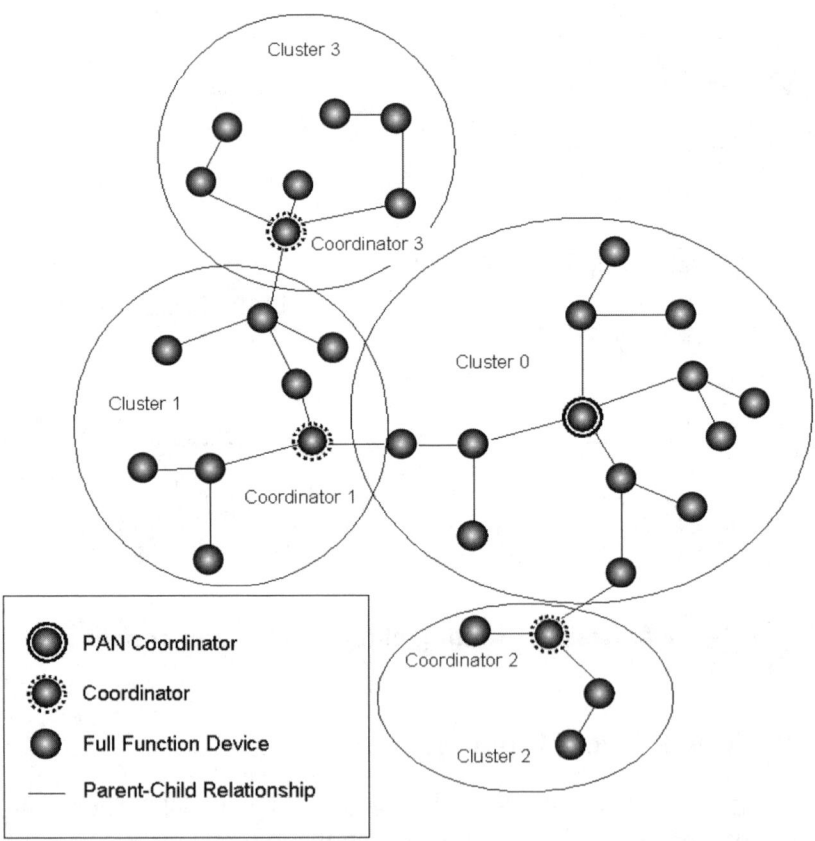

Figure 6–9: The cluster-tree network

The logical, short address of a network device in a cluster-tree network is now a hierarchical address consisting of two parts, a cluster identifier and a network device identifier. The network device identifier of a coordinator is always zero; the cluster identifier of the PAN coordinator is always zero.

In the cluster-tree network, the PAN coordinator may still transmit network status request messages. One way to do this is to have the network status request messages route, as in the cluster network, to the ends of the branches (the "leaves" of the tree). The network status response message is also handled in the same way, except that the device identifiers in the message are deleted by the eavesdropping coordinators; i.e., network status response messages leaving the coordinators contain only the identity of the cluster. This takes advantage of the address hierarchy to reduce the size of the network status response message, which, in a large network, may otherwise be impractical.

Note that, as in the cluster network, information extracted from the routing table of each network device may be included in the network status response messages to exploit routing redundancies.

Routing

A possible routing algorithm for a cluster-tree network is shown in Figure 6–10. In this algorithm, a device first checks to see if the cluster identifier of the destination is in its routing table. If not, the destination is an unknown cluster and the device routes the message to its parent.

However, if there is a cluster identifier match in the routing table, the device then checks those table entries with a cluster identifier match for a network identifier match. If one is found, the message is routed according to the table entry. If one is not found, the destination is an unknown network device in a known cluster, and the device must choose which network device to use to relay the message to the cluster by some set of arbitration rules.

If the cluster identifier of the destination is the same as that of the network device attempting to route the message, as in the cluster network, the message should be routed to the parent of the network device. However, the arbitration rules governing the relay of messages to other clusters can be significantly more flexible, and may consider, for example, the time a routing table entry was made—routing the message via the network device most recently added to the table, in the belief that that route is the most up-to-date and most accurately reflects the present state of the network. Other considerations are possible, many of which are optimal for particular applications.

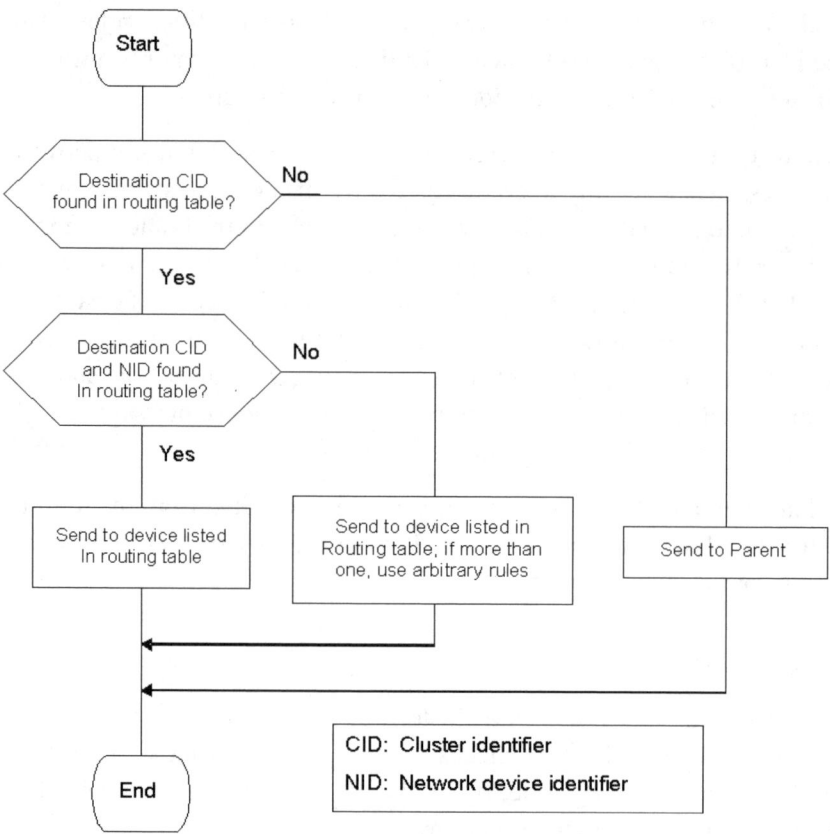

Figure 6–10:   A cluster-tree routing algorithm

THE NETWORK TOPOLOGY DECISION

Because IEEE Std 802.15.4 is capable of supporting a large number of ad hoc network topologies, an application designer employing IEEE Std 802.15.4 must select an appropriate network topology for the intended application. Further, if the selected network topology is a peer-to-peer topology in which multi-hop communication is planned, the routing algorithm must also be selected.

As an aid to this decision-making process, Table 6–1 summarizes the major network types discussed in this chapter, including their strengths, weaknesses, and potential applications. Note that Table 6–1 is only a guide, intending to establish general trends; almost any of these networks may be used successfully in almost any application, albeit with variations in network performance parameters such as message throughput and latency, and network device performance parameters such as power consumption.

The network types listed in this section are only examples; many other topologies and routing algorithms are possible.

Table 6–1: Network topology comparison

Network Type		Strengths	Weaknesses	Possible Applications
Star		Low message latency; centralized network control	Can cover only a limited physical area (single-hop communication)	Home automation; PC peripherals
Peer-to-Peer		Can cover a large physical area (multi-hop communication)	Higher message latency	Wireless sensor networks; industrial control and monitoring
	Flat	Simple network devices	Does not scale well as the number of potential destination devices increases	Wireless sensor networks
	Cluster	Supports a larger number of potential destination devices	Uneven power consumption among network devices	HVAC systems
	Cluster Tree	Can support very large networks	Network maintenance overhead	Industrial control and monitoring

Part III

Chapter 7 System Design Considerations

...a systemic view

IEEE Std 802.15.4 was created to address applications that could not be satisfied with existing WLAN and WPAN communication protocols, primarily due to the cost, size, and/or current drain of their implementations. This chapter discusses system design considerations needed to take advantage of the capabilities of IEEE Std 802.15.4 and achieve low cost, size, and current drain implementations of the standard.

To reduce cost, size, and current drain of its implementations, IEEE Std 802.15.4 was designed for a high level of integration, with a minimum number of external parts. Several features in the protocol help achieve this goal.

Direct Sequence Spread Spectrum

The use of Direct Sequence Spread Spectrum (DSSS) enables an IEEE Std 802.15.4 implementation to be largely a digital one, with only a relatively small number of analog circuits. This "future-proofs" the standard; expected improvements in integrated circuit lithography will increase digital circuit density and should lead to ever lower implementation costs. DSSS offers other implementation advantages. The DSSS processing gain provides rejection to interfering signals, reducing the requirements on the channel filter. The ability to easily design orthogonal, multi-level signals with DSSS enables the IEEE Std 802.15.4 2.4 GHz physical layer to have simultaneously a relatively fast data rate of 250 kb/s (to finish transmission and return to sleep quickly) and a relatively low symbol rate of 62.5 kSymbols/s (to minimize current drain while active); both of these features improve average power consumption. Further, in some regulatory jurisdictions, some type of spreading is required, and DSSS enables faster network discovery and synchronization than does frequency hopping spread spectrum (FHSS). For these reasons, DSSS was chosen over FHSS and narrowband techniques.

High channel spacing/modulation bandwidth ratio

IEEE Std 802.15.4 has a high ratio of channel spacing to modulation bandwidth. That is to say, adjacent channels are far away from the desired signal, compared to the width of the desired signal (see Figure 7–1). The first null in the transmitted

IEEE Std 802.15.4 spectrum in the 915 MHz band is 600 kHz away from the signal center, with 1.4 MHz spacing to the edge of the adjacent channel signal; in the 2.4 GHz band the first null is 1.5 MHz away from the signal center, with 3.5 MHz spacing. In both cases, the spacing-to-bandwidth ratio is $1.4/0.6 = 3.5/1.5 = 2.33$. However, IEEE Std 802.15.4 requires no selectivity to the adjacent channel, only the alternate channel, resulting in an effective ratio of $3.4/0.6 = 8.5/1.5 = 5.67$. In comparison, the transmitted IEEE Std 802.11b spectrum, in which the first null is 11 MHz away, with 14 MHz to the edge of the next nonoverlapping adjacent channel, has a ratio of only $14/11 = 1.27$; IEEE Std 802.11b requires 35 dB of selectivity to this channel. IEEE Std 802.15.4 channel filter stopbands are therefore well removed from the desired passband.

This has several beneficial effects:

- Receiver types that employ integrated selectivity, such as low-IF and zero-IF architectures, are possible. Discrete surface acoustic wave (SAW), ceramic, or crystal filters are not needed. This reduces implementation size and cost.
- For low-IF and zero-IF receiver implementations, the low-pass corner frequency of the channel filter may be raised. This reduces the size and cost of the integrated filter.
- Fewer poles of channel filtering may be used. This also reduces the size and cost of the filter.
- With fewer poles of filtering, and a higher corner frequency, the time required to settle turn-on transients in the channel filter may be reduced. This enables a faster receiver warm-up, minimizing battery power lost during the warm-up period.
- The relatively large channel spacing means that a high synthesizer reference frequency may be used. Because synthesizer lock time is related to the frequency at which the synthesizer's phase detector operates, a higher reference frequency enables shorter synthesizer lock times, which allows shorter warm-up times and therefore improves battery life.

Relaxed transmitted error vector magnitude (EVM) requirements

The relaxed receive selectivity and transmitted error vector magnitude (EVM) requirements of IEEE Std 802.15.4 allow the sideband noise specifications of the receiver local oscillator and transmitter oscillator to be relaxed, perhaps to the point that a very low-cost, fully-integrated oscillator may be employed. This integrated oscillator, which could be of the inductor-capacitor (LC) or (preferably) the resistor-capacitor (RC) type, would eliminate the large and costly external voltage controlled oscillator (VCO) resonator, and other associated passive components, that would otherwise be needed.

Figure 7–1: Comparison of IEEE Std 802.11 and IEEE Std 802.15.4 2.4 GHz PHY selectivity requirements

Constant-envelope modulation

A major factor leading to reduced transmit power consumption is the use of constant-envelope modulation, in the form of half-sine-shaped O-QPSK in the high band and raised-cosine-shaped BPSK in the low band. Transmitters must be designed, of course, to transmit the peak envelope power required by their modulation format; for simple circuits suitable for IEEE Std 802.15.4 implementations this requires the amplifier dc bias power (which can be a significant fraction of the total transmitter power consumption) to be proportional to the peak envelope power. The *average* envelope power, on the other hand, is the significant parameter when determining range (given a fixed data rate). If a modulation format is chosen such that the peak envelope power is greater than the average envelope power, dc bias power must be spent in the transmitter commensurate with the higher peak power, even though the range achieved is that of the lower average power—clearly an undesirable situation for power-sensitive applications. (Such modulation formats, however, may have advantages, such as spectral efficiency, that make them suitable for other applications.) IEEE Std 802.15.4 therefore employs a modulation format having a constant envelope, in which the peak and average envelope powers are equal (a peak- to average-power ratio of one). An implementation of IEEE Std 802.15.4 may take advantage of this to employ a simple and relatively efficient transmitter.

Relaxed receiver blocking specification

The IEEE Std 802.15.4 receiver blocking specification, −20 dBm, is significantly less than that of other WLAN and WPAN standards, and it enables an IEEE Std 802.15.4 system designer to reduce the current drain of the receiver front end accordingly.

TIME AND FREQUENCY REFERENCE TRADEOFFS

The design of the time and frequency reference subsystem of an IEEE Std 802.15.4 implementation can have a marked effect on overall system cost, size, and current drain. One possible approach to the design of a stand-alone IEEE Std 802.15.4 implementation employs two reference crystal oscillators, a high-frequency oscillator for the RF frequency reference and a low-frequency oscillator for the protocol time base. In this design, the low-frequency (and therefore lower power) crystal oscillator for the time base runs constantly, providing protocol timing information. The high-frequency (and therefore higher-power) crystal oscillator for the RF frequency reference is then started immediately prior to a transmit or receive period and is turned off immediately afterward, thereby minimizing its power consumption. If the required warm-up time of the RF frequency reference is low enough so that it is not a significant factor in total energy consumption for the application under consideration, this approach can produce low average power consumption, at the cost of requiring two external crystals.

A second approach employs only a single, high-frequency crystal oscillator. In this scenario, the oscillator is used for both the RF frequency reference and the protocol time base, and it is therefore never turned off. (The oscillator output is divided down to a low frequency to be used as the protocol time base.) When needed as the RF frequency reference, its output is used directly; because it is already operating, no warm-up period is needed and therefore no power is lost there. This second approach is lower cost than the first, because only one crystal is needed; however, because a higher-frequency oscillator, plus frequency division circuitry, is used for the time base, the average power consumption may be higher. This trade must be evaluated by the system designer based on the expected applications of IEEE Std 802.15.4 implementation.

SINGLE-CHIP VERSUS MULTIPLE-CHIP IMPLEMENTATION

An integrated circuit (component) manufacturer entering the IEEE Std 802.15.4 market is faced with the decision of whether its design should be a two-chip design, in which IEEE Std 802.15.4 RF transceiver is on a separate chip from the protocol handler (Figure 7–2a), or a one-chip design, in which the RF transceiver and protocol handler are integrated on the same die (Figure 7–2b). A one-chip design, because it has the highest level of integration and fewest number of IC packages, may have the smallest implementation and lowest component cost, and it may be optimum for stand-alone applications that do not have an available host processor (as described below). However, a two-chip design may employ more optimal IC processes that are better suited to the circuit requirements (e.g., BiCMOS for the RF transceiver and CMOS for the protocol handler), and it could result in a reduced time to market compared to a chip in a compromise integrated circuit (IC) process chosen for the one-chip design. In addition, if a wide product line of stand-alone IEEE Std 802.15.4-enabled products is envisioned, a two-chip design may allow the substitution of multiple processor types as the protocol handler IC, having different amounts of memory and capable of economically supporting a range of applications of varying complexity. The two-chip design also allows flexibility in the choice of RF transceiver with a single protocol handler; for example, one may want to enable either a dedicated IEEE Std 802.15.4 RF transceiver or one capable of servicing multiple protocols, for example both IEEE Std 802.15.4 and IEEE Std 802.11a. A compromise alternative to this decision is to employ two IC die, but placed in a single IC package.

OEM Implementations

The system design decisions faced by an original equipment manufacturer (OEM) are equally interesting. One scenario involves the addition of IEEE Std 802.15.4 capability to a larger system; for example, when a general-purpose host processor is available to run an IEEE Std 802.15.4 protocol stack, and a dedicated IEEE Std 802.15.4 protocol-handling processor is not used. In this case, shown in Figure 7–2c, the cost of the dedicated processor may be saved by employing a smaller chip performing only IEEE Std 802.15.4 RF transceiver functions—half of the "two-chip" design just described. However, software integration is made more difficult, especially if the host processor is running other real-time applications, because the performance of many protocol functions cannot be delayed. Interrupt handling and context switching, for example, must be carefully evaluated to ensure that enough processing capability [e.g., million instructions per second (MIPS), random access memory (RAM)] is always available, and the throughput and latency performance of the communication link between the IEEE Std 802.15.4 transceiver and host processor must be understood. Often the worst-case scenario, when the host processor is most heavily loaded, is difficult to predict and test, especially for embedded systems. If the software integration problem becomes untenable, a one-chip IEEE Std 802.15.4 implementation may be used to off-load the host processor, as shown in Figure 7–2d.

Lightly loaded host processors, on the other hand, may have little difficulty meeting the real-time timing requirements of the protocol stack. However, unless specifically designed for the task, the host processor may not perform the bursty computation requirements of the IEEE Std 802.15.4 protocol in a power-efficient manner; the average power consumption of the resulting IEEE Std 802.15.4 implementation may suffer as a result. The system designer may achieve the best compromise between implementation cost and power consumption by judiciously placing only those stack components on the host processor that best fit the needs of the intended application.

IEEE Std 802.15.4 was designed to be compatible with low-cost 8051- or HC08-based microcomputers operating at a bus speed of only a few megahertz. This design involved a trade between the utility and flexibility of multiple protocol layers, and the software size reduction that occurs when the protocol is defined as a single layer. The compromise reached was to keep the layered architecture, to retain compatibility with other IEEE 802 standards, but simplify each protocol layer as much as possible.

TIME AND POWER MANAGEMENT

IEEE Std 802.15.4 is designed to support a time base with a tolerance as great as +/–40 ppm. This allows the use of very inexpensive reference crystals, minimizing cost. At 2.4 GHz, the standard also supports a beacon interval of $2^{14} * 0.01536$ s = 251.65824 s, or over four minutes; this, combined with a minimum beacon PPDU length of 544 µs, enables a transmit duty cycle of $(544 * 10^{-6})/ 251.65824 = 2.16 * 10^{-6}$. Lower values are, of course, possible if beacons are not used at all.

However, a *receiving* device must be receiving at the time the beacon is sent, a time that is uncertain due to the imperfect accuracy of both transmit and receive time bases (Figure 7–3). Without time base regulation, the receiver must receive for a period of $2\varepsilon T_{beacon} + T_c$ s every T_{beacon} s, where ε is the sum of the transmit time base tolerance ε_{Tx} and receive time base tolerance ε_{Rx}; the lowest possible average duty cycle of a receiving device is therefore

$$2\varepsilon + \frac{T_c}{T_{beacon}}.$$

It is clear that, no matter how long the beacon period T_{beacon} is made, the duty cycle is limited by the attainable time base stability ε. (Note that neither transmit

Figure 7–2: Diverse IEEE Std 802.15.4 hardware implementations

nor receive warmup times are included in this analysis.) For 40 ppm timebases at both transmitter and receiver, this effect becomes significant for $T_{beacon} > 0.5$ s or so; to achieve the very low duty cycles (and resulting long battery life) of which devices employing IEEE Std 802.15.4 are capable, an implementation may therefore require a more stable time base than that specified in the standard. Although the simplest way to do this may be to replace, for example, a 40 ppm crystal oscillator with one of greater stability, this would usually increase the cost of the time base significantly—especially if stabilities below 5 ppm were required. A lower cost approach may be developed by recognizing that the time variation between sequential beacons from the same device is likely to be highly correlated. It is then possible for the receiving device to reduce the device duty cycle by tracking the relative received beacon time from beacon to beacon, and therefore employing a reception window smaller than $2\varepsilon T_{beacon} + T_c$.

Alternatively, a time base drift specification, limiting the allowable time difference between successive beacon periods, could have been specified in IEEE Std 802.15.4, as has been done in other standards. It is clear that a drift specification is not needed for *macBeaconOrder* = 0; when T_{beacon} = 15.36 ms, any possible drift is insignificant when compared to the beacon PPDU length (544 μs at 2.4 GHz) plus warmup time (a total of at least, say, 600 μs). The "drift time," even assuming an 80 ppm difference between devices, does not become significant when compared to 600 μs until one reaches *macBeaconOrder* = 7 (1.96608 s), when it is 157 μs.

It is also clear that a drift specification is not useful for *macBeaconOrder* = 14; a device could move from hot to cold during the four-minute beacon period, and the warm-up time estimate could only be improved by improving the accuracy of the time base, an undesirable option. Very small, single-chip IEEE Std 802.15.4 implementations (perhaps with button cell batteries) may have very low thermal inertia; when moving between temperature extremes during a 16-second beacon period (*macBeaconOrder* = 10), for example, they could well vary by +/–40 ppm, so a drift specification could not be applied here, either, without adding a significant (i.e., costly) restriction on the hardware. In this regime, *macBeaconOrder* = 10 to 14, the receiving time must be specified based on the time base accuracy specification.

This leaves a small window in which a drift specification would be useful, *macBeaconOrder* = 7 to 9. It was decided that specifying drift only for three *macBeaconOrder* values was not appropriate for a simple protocol with low implementation cost, so a time base drift specification is not included in IEEE Std 802.15.4.

The discussion above is an example of the importance of time and power management in the design of IEEE Std 802.15.4 implementations, and the concern for them that went into the design of the standard. Note that, because the standard is designed to support duty cycles of 0.1% or less, the standby power consumption of the implementation can be the dominant factor in average power consumption (and therefore battery life). It is therefore rewarding to pay attention to standby power consumption during the design process. In addition, because active times are brief (as short as 544 μs for a minimum-length 136-bit beacon PPDU at 2.4 GHz), warm-up times can be a significant fraction of overall active time and can therefore affect the minimum obtainable duty cycle. This effect is important if maximum battery life is to be achieved.

Figure 7–3: Effect of imperfect transmitter and receiver time

If the application can support a star network configuration with a mains-supplied PAN coordinator, and especially if the PAN coordinator is to be the destination of all network traffic (for example, in a lighting network, in which several wireless switches control a lamp PAN coordinator), the other network devices may have an arbitrarily low duty cycle by operating the network without a beacon. In this mode, the PAN coordinator is receiving constantly; the network is in a constant contention access period.

The switches need not transmit or receive at all until some event (e.g., the toggling of a switch) causes a message transmission from a switch to the lamp. Because a lamp is typically connected to the mains, having a near-100% receive duty cycle is not an onerous requirement in this application, and it enables the designer to craft a system in which the switch battery life may be exceptionally long.

ANTENNAS

Because many IEEE Std 802.15.4 products are expected to be physically small, and very sensitive to product cost and power consumption, antennas can be very important to a successful design.

The physically small antenna required in many applications may be inefficient unless careful attention is paid to the design of both the antenna and its placement in the product. An inefficient antenna can significantly reduce both the transmit and receive range, something that can be compensated for only by increasing transmit power and receiver sensitivity—both options that can significantly reduce battery life.

There is a trade to be made among the physical size, instantaneous fractional bandwidth BW, and the maximum achievable radiating efficiency η of electrically small antennas (i.e., those with maximum physical dimensions significantly less than their operating wavelength λ). Specifically [21],

$$\eta = \frac{2\left(\frac{2\pi r}{\lambda}\right)^3}{BW},$$

where $2r$ is the maximum dimension of the electrically small antenna (i.e., the diameter of the smallest sphere that can completely contain the antenna). From this equation, two important relationships may be noted:

- The antenna efficiency is proportional to the ratio of the antenna size and the operating wavelength, *cubed*. This means that, all else equal—including the physical size of the antenna—operation at a higher frequency (a lower wavelength λ) improves antenna efficiency, as does increasing the size of the antenna while keeping the frequency of operation constant. If, however, the antenna size scales with wavelength (as is the case for a half-wave dipole, for example), its efficiency is unaffected by the frequency of operation.

- Attainable fractional bandwidth BW is inversely proportional to attainable efficiency η. In most applications, the efficiency of even well-designed small antennas for any of IEEE Std 802.15.4 bands is low enough that the bandwidth required is easily achieved.

In addition to the antenna size $2r$, it must be recalled that strong circulating (i.e., non-radiating) fields exist near an antenna. These fields vary as d^{-2} and d^{-3}, where d is the distance from the antenna. The fields are therefore very strong close to the antenna, and significant out to a radius of at least $\lambda/10$. Any material in this region, including batteries, circuits, displays, and so on, produces additional loss that can significantly affect antenna efficiency. One therefore should plan on a "keep out" region in the product design of 1.2 cm or so for 2.4 GHz operation, and approximately 3.5 cm for 868 MHz operation; the more material in this region, the lower the antenna efficiency will be. When keep out regions of this size are not possible, the increased loss should be considered in the implementation design. Note that the effect of nearby materials can counteract the increased efficiency of a larger antenna over a smaller one, if the larger antenna must be placed near a lossy object while the smaller antenna is kept away from it.

This theory must be matched with the practical realities of antenna design for small, low-cost products. Because any volume taken by an antenna inside a product is not available for other uses (the placement of circuits, batteries, etc.), it is nearly always desirable to minimize it. Several factors affect this product design decision:

- *The frequency of operation.* As stated above, for constant efficiency the antenna size scales with the operation wavelength; operation in the 2.4 GHz band would therefore consume less antenna volume.

- *Products for multiple bands.* If a single product design is intended to be used for both 868 MHz and 2.4 GHz operation, for example, space must be made available for the larger 868 MHz antenna.

- *Internal noise sources.* Due to their proximity to the antenna in a small product, even relatively weak noise sources can play havoc with a wireless transceiver. Typical sources are digital circuits (especially the busses of microprocessors and microcomputers, and serial data interfaces) and switching power converters. These components, and the circuit board traces associated with them, should be placed as far as possible from the antenna, something that becomes more difficult as the antenna size increases. When possible, of course, it is desirable to arrange operation of the digital circuits so that they are not clocked when the transceiver is active.

- *Use of external antennas.* In general, external antennas (those outside the product housing) perform better electrically; they are typically away from lossy materials and have fewer size restrictions, so they may be larger. If a directional antenna is needed (perhaps for a long-range application), an external antenna is often the best choice; they may also be required in applications such as wireless home appliances (e.g., a refrigerator) in which the radio transceiver is enclosed in a metal housing. External antennas also maximize the internal volume available for other product components. However, they may cost more than internal antennas (especially if RF connectors are needed) and, because they lack the protection of the product housing, they may be prone to mechanical failure due to environmental factors, handling, and so on. The loss of the feedline between the antenna and transceiver must also be considered in the product design.

If an internal antenna is the decision, there are several design choices available. The first choice is often to employ a circuit board trace as an antenna. This is certainly the cheapest approach, because no antenna components need be purchased, handled, or placed on the board, but it is not free, because circuit board area must still be purchased for it. On the negative side, circuit board antennas tend to have significant loss, due to their resistance (caused by the thinness of the copper trace) and their proximity to lossy materials such as the circuit board material itself, nearby ground planes and traces, and other circuit components.

Wire antennas, such as loops and dipoles, allow the antenna to be placed above the plane of the circuit board, improving its efficiency, and they are often a good compromise between the performance of external antennas and the cost of circuit board trace antennas. Wire antennas may require tuning for best performance, due to mechanical shape and size variations that affect their frequency of resonance, and they may need non-metallic physical supports in certain applications.

Specialized antenna components, composed of ceramic materials, are also available. They offer a smaller physical size than that attainable with wire antennas and do not require tuning; however, they are more expensive than wire antennas.

PRODUCT DESIGN FLEXIBILITY

IEEE Std 802.15.4 is designed to have the flexibility to serve in a wide variety of applications—from personal computer peripherals and toys to home automation and industrial controls. These applications, however, may have very different cost, size, and current drain goals, and the ideal design for one may be unsuitable for another. Figure 7–4 is a spider chart showing the relative importance of size, cost, and current drain for three possible IEEE Std 802.15.4 applications. Note that these determinations are exemplary, and will vary greatly, depending on implementation details and the specific applications chosen within these broad categories.

In this illustrative example, cost is of primary importance in the home automation application, whereas current drain is of the least importance—possibly because the network may have access to mains power. Similarly, current drain is of primary importance in the industrial control application, whereas size is of the least importance—possibly because the application is a large industrial motor controller. Finally, size is the primary consideration in the PC peripheral application—for example, a wireless mouse or joystick—whereas cost is of the least importance. The application designer must make the right trades in these three dimensions to optimize an IEEE Std 802.15.4 product.

Figure 7–4: Example of relative importance of performance parameters for different applications

Chapter 8   Real-World Issues

...a touch of reality

Users, implementers, and system integrators of IEEE Std 802.15.4 should be aware of some factors that may affect the performance of their systems resulting from interaction with the "real world." Two of these factors are the coexistence of IEEE Std 802.15.4 products with other users of the radio frequency spectrum, and the interfacing of different communication protocols.

COEXISTENCE

Because the bands used by the two IEEE Std 802.15.4 PHY layers are unlicensed in nearly all countries, there are devices employing other communication protocols sharing the bands. In the 868 MHz band, for example, there are proprietary low-data-rate protocols; in the 915 MHz band, proprietary protocols and cordless telephones; and in the 2.4 GHz band, cordless telephones, microwave ovens, WLANs, and other WPANs. In most regulatory regions, devices operating in these unlicensed bands must accept interference to them caused by other services operating in the same band, and their users do not have legal redress for any performance degradation caused by the other services. Although this may be the regulatory reality, it is not the market reality; users purchasing new devices reasonably expect that not only will the new devices operate as desired, but that they also will not disrupt the operation of existing products they may have. A device not meeting these two criteria may have a difficult time remaining on the market. Coexistence is therefore an important economic issue.

Several features of IEEE Std 802.15.4 were designed with coexistence with other services in mind. Although the issue can be considered as two problems (the protection of other services from interference caused by IEEE Std 802.15.4 transmissions, and the protection of IEEE Std 802.15.4 implementations from interference caused by the transmissions of other services), in practice most solutions to one of these problems are also solutions to the other. Some features of IEEE Std 802.15.4 that reduce potential interference to other services are listed below; much of this list is derived from the Coexistence Annex of the standard.

- *Low Transmitted Power*: Although radio regulations in many parts of the world (including Part 15.247 of the FCC rules [22]) allow operation in the

915 MHz and 2.4 GHz bands with relatively high output power (as much as 1 W in some cases), the typical IEEE Std 802.15.4 device will likely operate with much lower transmit power. A key metric of IEEE Std 802.15.4 is cost, and achieving greater than 10 dBm transmit power in a low-cost system-on-chip (SOC), although feasible, will be economically disadvantageous. Furthermore, FCC and European regulations (ETSI EN 300 220-1 V1.3.1 [5], CEPT Recommendation 70-03 [4], and ETSI SN 300-328 [6]) for out-of-band emissions make it difficult to employ powers above 10 dBm without additional, expensive filtering of IEEE Std 802.15.4 transmitted spectrum. These factors, plus the IEEE Std 802.15.4 requirement that devices be capable of transmitting −3 dBm, will limit the distribution of devices with greater than 10 dBm transmit power to a few specialized applications.

- *Channel Alignment:* A concern in many applications of IEEE Std 802.15.4 that employ the 2.4 GHz PHY is coexistence with WLANs using the popular IEEE Std 802.11, particularly those WLAN installations in which the maximum of three non-overlapping WLAN channels are employed. Figure 8–1 shows the alignment between the North American non-overlapping IEEE Std 802.11b channels and IEEE Std 802.15.4 channels. There are four IEEE Std 802.15.4 channels that fall in the guard bands between (or above) the three IEEE Std 802.11b channels (n =15, 20, 25, 26 for North America, n = 15, 16, 21, 22 for the European non-overlapping WLAN channels). Although the energy in this guard space will not be zero, it will be lower than the energy within the channels, and operating an IEEE Std 802.15.4 network on one of these channels will minimize interference between systems. When performing dynamic channel selection (controlled by the upper layers), either at network initialization or in response to channel impairment, an IEEE Std 802.15.4 device will scan a set of channels specified by the "ChannelList" parameter in the MLME-SCAN.request primitive. For IEEE Std 802.15.4 networks that are installed in areas known to have high IEEE Std 802.11b activity, the Channel-List parameter can be defined as one of the above sets (either [15, 20, 25, 26] or [15, 16, 21, 22], as appropriate) in order to maximize the coexistence of the networks.

- *Channel Selection:* Prior to network formation, the candidate PAN coordinator performs a scan of all channels in its channel list, to identify IEEE Std 802.15.4 PANs already in existence. If a suitable PAN is identified, at the discretion of the network layer the candidate may choose to join that PAN, rather than create a second PAN. This behavior will minimize the number of PANs existing in a band, thereby reducing potential interference to other services. Should no acceptable PANs be identified, however, the candidate PAN coordinator may choose to start a new PAN. It selects an unoccupied

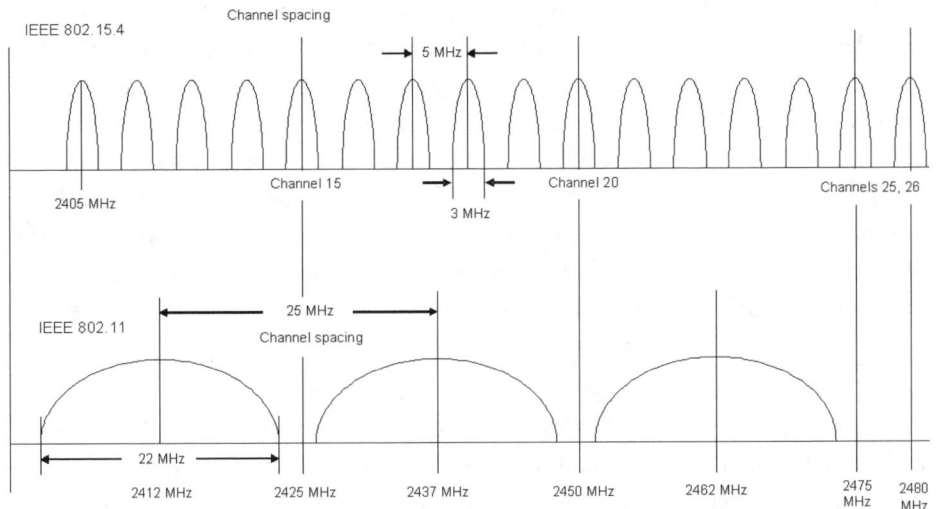

Figure 8–1: Alignment of the North American IEEE Std 802.11 non-overlapping channel plan and IEEE Std 802.15.4 2.4 GHz PHY channels

channel for the new PAN from the list of channels in the ChannelList parameter (which may be determined by an energy detection channel scan), a second behavior that avoids transmission on frequencies occupied by other services.

Should interference appear on the channel occupied by the PAN, the upper layers of the PAN coordinator execute a dynamic channel selection algorithm (not defined in IEEE Std 802.15.4). The PAN coordinator scans the available channels in its channel list and selects a new channel for the PAN. It then returns to the original, impaired channel and makes a broadcast message to the PAN, identifying the new channel to which the PAN is to move. This procedure helps IEEE Std 802.15.4 systems avoid causing interference to, and receiving interference from, other services.

- *Clear Channel Assessment:* The CSMA-CA channel access mechanism in IEEE Std 802.15.4 performs a Clear Channel Assessment (CCA) prior to transmission. The PHY requires that at least one of three CCA methods be used—energy detection (ED) over a certain threshold, detection of a signal with IEEE Std 802.15.4 characteristics, or a combination of the two. Use of the energy detection option improves the coexistence behavior of IEEE Std 802.15.4 with other services by allowing transmission backoff if the channel is occupied by any device, regardless of the communication protocol it may use.

- *Use of Spread Spectrum Techniques:* The direct sequence spreading technique used by IEEE Std 802.15.4 provides some protection for users of narrowband (i.e., less than 200 kHz wide) communication protocols, by spreading the 250 kb/s data transmission over a bandwidth greater than 2 MHz. This reduces the energy of IEEE Std 802.15.4 transmission present in any given narrowband channel, protecting these services. Interestingly, even though its signal is spread, due to its low data rate IEEE Std 802.15.4 transmission bandwidth is comparable to that of IEEE Std 802.15.1 (Bluetooth); as a result an IEEE Std 802.15.4 PAN interferes with a Bluetooth PAN much as a second Bluetooth PAN would—affecting only three of Bluetooth's 79 hops.
- *Link Quality Indication:* IEEE Std 802.15.4 PHY layer standard specifies a Link Quality Indication (LQI). The LQI is performed on each received packet, and it may be implemented using received signal strength, a signal-to-noise ratio estimation, or a combination of these. LQI can be used to detect channel impairment caused by interference on a packet-by-packet basis, providing "real time" information to the upper layers of the channel condition, enabling the device to make an informed decision in dynamic channel selection.

IMPLEMENTATION OF COLOCATED TRANSCEIVERS

An implementer of a system employing two RF transceivers, such as a cell phone incorporating a WPAN, should be aware of interactions that may occur between the two devices. Many applications may require such a design, including those in which the cell phone incorporates a PAN coordinator and is used as the gateway device between the WPAN and the Internet. These interactions can be due to both radiated and conducted effects.

A primary concern when placing two RF transceivers in the same product (or, indeed, in close proximity) is the effect the radiated transmitted noise of one device may have on the receiver of the second device. All transmitters transmit, in addition to the desired signal, some amount of wideband noise, often for many hundreds of megahertz around the desired signal. Ordinarily, because the amplitude of this noise is so low, it is of no consequence, and it is not economically justifiable to insert filters or other circuits to reduce it further. However, if a receiver (or, more specifically, a receiving antenna) is brought very close to the transmitting antenna, the noise can become significant. In this case, the noise may affect the apparent receiving sensitivity of the second device, even though the frequencies of operation of the two transceivers differ greatly. If the communication protocols of the two transceivers are independent, so that the transmission of one device cannot be prohibited while the second device is receiving, special measures, in the form of adding selectivity to the transmitter output, may have to be performed to enable successful operation of the two transceivers.

A second radiated effect is the effect of receiver blocking. Blocking, a strong signal effect (as opposed to a noise effect), occurs when a very strong undesired signal enters a receiver and affects the dc bias of its circuits. This typically causes a loss of gain, leading to a loss of sensitivity. In a sense, this affects the "signal" portion of the receiver detector's signal-to-noise ratio, whereas wideband transmitted noise affects the "noise" portion. Blocking can happen, for example, if a 600-mW cellular transmitter is co-located with a WPAN receiver that is biased for low-power operation. Blocking performance can be improved by adding selectivity before the receiver input, if the offending signal is outside the passband of the victim receiver or, usually less desirable, by increasing the dc bias of the receiver circuits (i.e., increasing power consumption).

A conducted effect that can arise when employing two transceivers results from the sharing of a power supply. In portable devices this is often unavoidable, because the primary supply is a battery and there is rarely room in the size or weight budget for a second battery in the product. However, if one transceiver draws significant current from the shared supply (usually while transmitting), any voltage drop resulting from either internal or external resistance associated with that supply could affect the second transceiver. The textbook example of this phenomenon is the supply voltage modulation caused by a time division multiple access (TDMA) cellular transmitter, such as a global systems for mobile (GSM) transmitter, which, because it is at an audio rate, caused an audible tone in early GSM handsets. This problem can be corrected with suitable supply design, but the addition of a separate voltage regulator for the victim device is often required.

Chapter 9 Concluding Remarks

... summarizing

IEEE Std 802.15.4 was designed to meet the needs of an application space that has received little attention until now—that of wireless communication applications requiring only moderate data throughput, but needing very low power consumption and very low implementation cost. The IEEE Std 802.15.4 Task Group has worked to produce a standard that is at once flexible enough to meet the demands of a wide range of applications—from the wireless sensor networks used in intelligent agriculture and military applications, to the industrial control and monitoring networks, to consumer electronics and home automation uses—while retaining a minimum of costly options and rarely-used special features that, while optimizing performance in one application, add complexity to all others. The Task Group believes that, by striking the proper balance between specialization and complexity, it has maximized the utility of the standard to the wireless communication industry.

To maximize worldwide utility, IEEE Std 802.15.4 offers two physical layers: the lower PHY, including both the European 868.0–868.6 MHz and North American 902–928 MHz bands, and the upper PHY, in the more widely available 2.4000–2.4835 GHz band. At 868.3 MHz a single channel, with a raw data rate of 20 kb/s, is available; in the 902–928 MHz band, 10 channels, each with a raw data rate of 40 kb/s, are available; and in the 2.4000–2.4835 GHz band, 16 channels, each with a raw data rate of 250 kb/s, are available. Direct Sequence Spread Spectrum (DSSS) is employed in both physical layers to take advantage of the low cost, largely digital implementations that result—and to enable the standard to enjoy future cost reductions due to Moore's Law. In the lower PHY, a conventional, binary BPSK form of DSSS is employed, whereas in the upper PHY a form of orthogonal coding is employed with half-sine-shaped O-QPSK.

The combination of the relatively low transmitted power expected to be used by most compliant devices, the use of DSSS, the use of CCA techniques prior to network formation, and the low duty cycle of the MAC, should make IEEE Std 802.15.4 devices among the best of neighbors in the coexistence with other wireless technologies.

To achieve the IEEE Std 802.15.4 goals of wide utility without undue complexity, the MAC employs a simple slotted superframe with optional beacons. To support applications requiring guaranteed bandwidth, such as wireless keyboards and mice, time can be reserved in each superframe for specified network devices in guaranteed time slots (GTSs). The superframe time not occupied by beacons or GTSs can be designated as a contention period, in which devices may contend for the channel using a conventional CSMA technique. Contention-based access was chosen over polling-based access to reduce power consumption of the networked devices (by reducing their duty cycle), and to support multiple network topologies with a minimum of complexity.

The standard also specifies security based on the well-respected AES-128 encryption algorithm. Both message encryption and integrity checking are supported; for high-security applications requiring message integrity, the full 128-bit Message Integrity Code (MIC) may be added to each transmitted packet, whereas for applications able to trade some security in exchange for shorter message length the MIC may be truncated to 64 or 32 bits.

IEEE Std 802.15.4 supports both star and peer-to-peer networks, but like all IEEE 802 standards, it does not specify the network layer design. (However, it does specify network association and disassociation mechanisms.) The user may choose to employ a network specified by an industry consortium such as the ZigBee Alliance (http://www.zigbee.org), or design a specialized network that fits his application requirements.

This combination of worldwide availability, multiple medium access techniques, strong security, and multiple supported network architectures enables a wide variety of applications, and it is the driving force behind the success of IEEE Std 802.15.4. As more and more individuals and organizations discover the value of Low-Rate Wireless Personal Area Networking to improve their productivity and standard of living, this trend will continue.

References

[1] Braley, Richard C., Gifford, Ian C., and Heile, Robert F., "Wireless Personal Area Networks: An overview of the IEEE P802.15 working group," *ACM Mobile Computing and Communications Review*, vol. 4, no. 1, January 2000, pp. 26–34.

[2] Callaway, Ed, Bahl, Venkat, Gorday, Paul, Gutiérrez, José A., Hester, Lance, Naeve, Marco, and Heile, Robert, "Home Networking with IEEE 802.15.4, a Developing Standard for Low-Rate Wireless Personal Area Networks," *IEEE Communications Magazine*, special issue on Home Networking, vol. 40, no. 8, August 2002, pp. 70–77.

[3] Ennis, G., "Impact of Bluetooth on 802.11 direct sequence," *IEEE P802.11-98/319*, 1998.

[4] European Conference of Postal and Telecommunications Administration (CEPT), European Radiocommunications Committee; Relating to the Use of Short Range Devices (SRD). CEPT/ERC/REC Recommendation 70–03. December 2002.

[5] European Telecommunication Standards Institute, Electromagnetic compatibility and Radio spectrum Matters (ERM); Short Range Devices (SRD); Radio equipment to be used in the 25 MHz to 1 000 MHz frequency range with power levels ranging up to 500 mW; Part 1: Technical characteristics and test methods. ETSI EN 300 220-1 V1.3.1. Sophia-Antipolis, France: European Telecommunication Standards Institute, September 2001.

[6] European Telecommunication Standards Institute, Electromagnetic Compatibility and Radio Spectrum Matters (ERM); Wideband Transmission Systems; Data Transmission Equipment Operating in the 2,4 GHz ISM Band and Using Spread Spectrum Modulation Techniques. ETSI EN 300-328. Sophia-Antipolis, France: European Telecommunication Standards Institute. 2001.

[7] Gutiérrez, José, Naeve, Marco, Callaway, Ed, Bourgeois, Monique, Mitter, Vinay, and Heile, Robert F., "IEEE 802.15.4—A Developing Standard for Low-Power, Low-Cost Wireless Personal Area Networks," *IEEE Network Magazine*, vol. 15 no. 5, September/October 2001, pp. 12–19.

[8] Heegard, Chris, Coffey, John (Seán) T., Gummadi, Srikanth, Murphy, Peter A., Provencio, Ron, Rossin, Eric J., Schrum, Sid, and Shoemake, Matthew B., "High-performance wireless ethernet," *IEEE Communications*, vol. 39, no. 11, November 2001, pp. 64–73.

[9] Howitt, I., "Bluetooth performance in the presence of 802.11b WLAN," *IEEE Transactions on Vehicular Technology*, vol. 51, 2002.

[10] Howitt, I., "WLAN and WPAN Coexistence in UL Band," *IEEE Transactions on Vehicular Technology*, vol. 50, 2001, pp. 1114-1124.

[11] Howitt, I., and Gutiérrez, J. A., "IEEE 802.15.4 low rate wireless personal area network coexistence issues," *Proceedings of the WCNC 2003*, 2003.

[12] Howitt, I., Mitter, V., and Gutiérrez, J., "Empirical study for IEEE 802.11 and Bluetooth interoperability," *IEEE Spring VTC 2001*, Rhodes, 2001.

[13] IEEE Std 802.11™-1999, IEEE Standard for Information technology—Telecommunications and information exchange between systems—Local and metropolitan area networks—Specific requirements—Part 11: Wireless LAN Medium Access Control (MAC) and Physical Layer (PHY) specifications.

[14] IEEE Std 802.15.1™-2002, IEEE Standard for Information technology—Telecommunications and information exchange between systems— Local and metropolitan area networks—Specific requirements—Part 15.1: Wireless Medium Access Control (MAC) and Physical Layer (PHY) Specifications for Wireless Personal Area Networks (WPANs).

[15] IEEE Std 802.15.4™-2003, IEEE Standard for Information technology—Telecommunications and information exchange between systems—Local and metropolitan area networks—Specific requirements—Part 15.4: Wireless Medium Access Control (MAC) and Physical Layer (PHY) Specifications for Low-Rate Wireless Personal Area Networks (WPANs).

[16] Karaouz, Jeyhan, "High-rate Wireless Personal Area Networks," *IEEE Communications*, vol. 39, no. 12, December 2001, pp. 96–102.

[17] Marks, Roger B., Gifford, Ian C., and O'Hara, Bob, "Standards in IEEE 802 unleash the wireless internet," *IEEE Microwave*, vol. 2, no. 2, June 2001, pp. 46– 56.

[18] O'Hara, Bob, Petrick, Al, *IEEE 802.11 Handbook: A Designer's Companion*. New York: Standards Information Network, IEEE Press, 1999.

[19] Poor, Robert Dunbar, Embedded networks: Pervasive, low-power, wireless connectivity, Ph.D dissertation, 2001, Massachusetts Institute of Technology, Cambridge, MA.

[20] Siep, Tom, *An IEEE Guide: How to Find What You Need in the Bluetooth*™ *Spec*. New York: Standards Information Network, IEEE Press, 2000.

References

[21] Siwiak, Kai, *Radiowave Propagation and Antennas for Personal Communications*, 2nd ed., Boston: Artech House.1998.

[22] U.S. Code of Federal Regulations, vol. 47, sec. 15.247. Washington, D.C.: U.S. Government Printing Office. 2001.

[23] U.S. Code of Federal Regulations, vol. 47, sec. 15.249. Washington, D.C.: U.S. Government Printing Office. 2001.

[24] U.S. Department of Commerce, National Institute of Standards and Technology, Specification for the Advanced Encryption Standard (AES). Federal Information Processing Standards Publication 197. November 26, 2001. http://csrc.nist.gov/publications/fips/fips197/fips-197.pdf.

[25] van Nee, Richard, "A new OFDM standard for high rate wireless LAN in the 5 GHz band," *Proc. IEEE Veh. Tech. Conf.*, 1999, vol. 1, pp. 258–262.

[26] Zimmermann, Hubert, "OSI reference model—The ISO model of architecture for Open Systems Interconnection," *IEEE Trans.Commun.*, vol. COM-28, no. 4, April 1980, pp. 425–432.

[27] Zyren, J., "Reliability of IEEE 802.11 Hi rate DSSS WLANs in a high density Bluetooth environment," *Bluetooth '99*, 1999.

Glossary

access control list: A table used by a device to determine which devices are authorized to perform a specific function.

ad hoc network: A wireless network composed of communicating devices without preexisting infrastructure. An ad hoc network is typically created in a spontaneous manner and is self-organizing and self-maintaining.

advanced encryption standard: The symmetric encryption algorithm specified by the U.S. government in Federal Information Processing Standard (FIPS) 197. (http://csrc.nist.gov/encryption/aes/.)

application layer: The top layer in the open system interconnection (OSI) data communication model, in which application functions are performed that directly affect the end user.

association: The service used to establish a device's membership in a wireless personal area network.

authentication: The service used to establish the identity of one device as a member of the set of devices authorized to communicate securely to other devices in the set.

authentic data: Data with a source verifiable through cryptographic means.

bandwidth: The difference between the highest and lowest frequencies employed in a communication channel. More informally, the largest amount of data that can be sent per unit time over a communication channel.

baseband: A communication channel through which constant-potential symbols (e.g., dc voltages) are sent, without frequency shifting (e.g., modulation of a high-frequency carrier); the band of frequencies occupied by the signal before it modulates the carrier.

baud rate: A measure of signaling speed equal to the number of discrete conditions or signal events per second; the number of symbols per second in a communication system.

bit rate: A measure of the amount of information per second in a communication system. Also, the number of binary symbols per second in a communication system.

Glossary

Bluetooth: A Wireless Personal Area Network specification sponsored by the Bluetooth Special Interest Group (http://www.bluetooth.org) and standardized as IEEE Std 802.15.1.

broadband: Having a large bandwidth when compared to a reference; especially, a modulated signal having a large bandwidth when compared to its associated baseband signal. More informally, a high bit rate suitable for multimedia applications.

clear channel assessment: An evaluation of the communication channel prior to a transmission to determine if the channel is occupied.

confidentiality: Assurance that communicated data remains private to the parties for whom the data are intended.

coordinator: A full function device (FFD) with network device functionality that is also capable of providing synchronization services through the transmission of beacons and the creation of a superframe structure. If a coordinator is the principal controller of a PAN, it is called the PAN coordinator.

coverage area: The area where two or more IEEE Std 802.15.4 units can exchange messages with acceptable quality and performance.

cyclic redundancy check (CRC): An error detection code transmitted with a block of data in order to detect corruption of the data.

data integrity: Assurance that received data have not been modified from its original form.

data link layer: The layer between the Physical and Network layers of the open system interconnection (OSI) reference model for communication between computer systems, providing the functional and procedural means to establish, maintain, and release data links between network entities.

disassociation: The service that removes an existing association.

frame: The format of aggregated bits from a MAC sublayer entity that are transmitted together in time.

full function device (FFD): A device capable of operating as a coordinator or network device, implementing the complete protocol set.

IEEE: *See:* **Institute of Electrical and Electronic Engineers**

IEEE Std 802.11: IEEE standard that specifies medium access and physical-layer specification for wireless connectivity among fixed, portable, and moving devices within a local area at data rates of 1 and 2 Mb/s.

IEEE Std 802.11a: Extension of the IEEE Std 802.11 using orthogonal frequency-division multiplexing (OFDM) modulation, operating in the 5GHz frequency band at data rates of up to 54 Mb/s.

IEEE Std 802.11b: Extension of the IEEE Std 802.11 standard using direct sequence spread spectrum modulation in the 2.4 GHz frequency band at data rates of up to 11 Mb/s.

industrial, scientific, and medical (ISM) bands: Radio-frequency bands reserved for these applications by international agreement. The ISM bands relevant to WPANs are located at 900 MHz, 2.4 GHz and 5.7 GHz.

Institute of Electrical and Electronic Engineers (IEEE): A non-profit, technical professional association with the goal of advancing global prosperity by fostering technological innovation, enabling members' careers, and promoting community world-wide. The IEEE promotes the engineering process of creating, developing, integrating, sharing, and applying knowledge about electro and information technologies and sciences for the benefit of humanity and the profession (http://www.ieee.org).

integrity code: A data string generated using a symmetric key that is typically appended to data in order to provide data integrity and source authentication (also called a message integrity code).

International Organization for Standardization (ISO): A worldwide, non-governmental federation of national standards bodies, promoting the development of standardization and related activities in the world with a view to facilitating the international exchange of goods and services, and to developing cooperation in the spheres of intellectual, scientific, technological and economic activity (http://www.iso.org).

International Telecommunications Union (ITU): An international organization within the United Nations system in which governments and the private sector coordinate global telecom networks and services (http://www.itu.int).

ISO: *See:* **International Organization for Standardization**

ITU: *See:* **International Telecommunications Union**

key establishment: A process by which two entities securely establish a symmetric key that is known only by the participating entities.

key management: Methods for controlling keying material throughout its life cycle, including creation, distribution, and destruction.

Glossary

key transport: A process by which an entity sends a key to another entity.

ISM bands: *See:* **industrial, scientific, and medical bands**

logical channel: One of a number of distinct channels on a physical communication link.

Logical Link Control sublayer: The upper portion of the Data Link Layer in the OSI reference model, responsible for the organization of data flow.

low-rate wireless personal area network (LR-WPAN): A wireless personal area network, optimized for low-data-rate applications emphasizing long battery life and low implementation cost.

MAC sublayer: *See:* **Medium Access Control sublayer**

MAC protocol data unit (MPDU): The unit of data exchanged between two Medium Access Control entities.

Medium Access Control sublayer (MAC sublayer): The lower portion of the Data Link Layer in the open system interconnection (OSI) reference model, responsible for acquiring the right to use the underlying physical communication medium.

message integrity code: *See:* **integrity code**

mobile device: A device that uses network communications while in motion.

narrowband: Having a small bandwidth when compared to a reference, especially, a modulated signal having a bandwidth comparable to, or less than, its associated baseband signal. More informally, a low bit rate suitable for voice and data applications.

network device: A reduced function device (RFD) or full function device (FFD) implementation containing an IEEE Std 802.15.4 medium access control and physical interface to the wireless medium.

network layer: The layer between the Data Link and Transport layers of the open system interconnection (OSI) reference model for communication between computer systems, providing routing and switching functions and procedures.

network topology: The logical structure of a network.

Open System Interconnection (OSI): The seven-layer ISO model of architecture for communication between (possibly heterogeneous) computer systems.

orphaned device: A device that has lost contact to its associated personal area network (PAN) coordinator.

packet: The format of aggregated bits that are transmitted together in time across the physical medium.

PAN coordinator: A coordinator that is the principal controller of a personal area network (PAN). An IEEE Std 802.15.4 network has exactly one PAN coordinator.

payload data: The contents of a data message that is being transmitted.

payload protection: The generic term for the provision of security services on payload data, including confidentiality, data integrity, and authentication.

peer-to-peer network: A substantially homogeneous network in which devices communicate as equals, often employing distributed, multi-hop routing protocols.

personal area network (PAN): A wireless network for the personal operating space.

personal operating space: The space about a person or object that is typically about 10 m in all directions and envelops the person or object whether stationary or in motion.

physical layer: The layer below the Data Link Layer of the open system interconnection (OSI) reference model for communication between computer systems, providing the physical connections between data link entities.

portable device: A device that may be moved from location to location, but only uses network communications while at a fixed location.

protocol data unit (PDU): The unit of data exchanged between two peer entities.

pseudo-random number generation: The process of generating a deterministic sequence of bits from a given seed that has the statistical properties of a random sequence of bits when the seed is not known.

random number generator: A device that provides a sequence of bits that is unpredictable.

reduced function device (RFD): A device operating with a minimal implementation of the IEEE Std 802.15.4 protocol that is capable only of functioning as a network device.

security suite: A group of security operations designed to provide security services on MAC frames.

service access point: Any entity that provides access to the service interface.

service data unit (SDU): Information that is delivered as a unit through a service access point.

service primitive: The basic unit of service provided between layers of a communication protocol.

star network: A network employing a single, central device through which all communication between devices must pass.

symmetric key: A secret key that is shared between two or more parties that may be used for encryption/decryption or integrity protection/integrity verification depending on its intended use.

transaction: The exchange of related, consecutive frames between two peer medium access control (MAC) entities, required for a successful transmission of a MAC command frame or a data frame.

transport layer: The layer between the Network and Session layers of the open system interconnection (OSI) reference model for communication between computer systems, providing transparent transfer of data between session entities.

wireless local area network (WLAN): A computer communication network spanning at most a campus-sized (~1 km) area, employing radio, infrared, or other wireless physical media.

wireless medium (WM): The medium used to implement the transfer of protocol data units (PDUs) between peer physical layer (PHY) entities of a low-rate wireless personal area network (LR-WPAN).

wireless metropolitan area network: A computer communication network spanning at most a city-sized area, employing radio, infrared, or other wireless physical media.

wireless personal area network (WPAN): A computer communication network spanning the personal operating space, employing radio, infrared, or other wireless physical media.

ZigBee Alliance: An association of companies working together to create a very low-cost, very low power consumption, two-way, wireless communications standard (http://zigbee.org).

Index

A

Access control
    See Security, Features, Access control
Access control list (ACL) mode
    See Security, Modes, Access control list (ACL) mode
Acknowledgment 30, 31, 56, 64, 65, 67, 75, 85, 91, 103
    ACK xv
    Frame 33, 34, 35, 59, 68, 70, 85–88
Ad hoc network 12, 60, 71, 102, 103, 105, 112, 143
Address 11, 85, 86, 89, 92, 102, 103, 107, 109
    64-bit 60, 71, 75
    Encrypted 91
    Extended 99
    IEEE 71
    MAC 67, 71, 85, 86, 87
    Short 60, 71, 75, 111
Adjacent channel 28, 51, 117, 118
Advanced Encryption Standard (AES)
    See Security, Advanced Encryption Standard (AES)
AES-CBC MAC-64
    See Security, Suites
AES-CBC-MAC-128
    See Security, Suites
AES-CBC-MAC-32
    See Security, Suites
AES-CCM-128
    See Security, Suites
AES-CCM-32
    See Security, Suites
AES-CCM-64 *See* security, suites
AES-CTR
    See Security, Suites
Alternate channel 51, 118
Ant-based algorithms 103–104
Antenna 48, 126–128
    Bandwidth 126
    Effect of frequency of operation 48, 127
    Effect of small size 39, 126, 127
    Effects of internal noise sources 127
    Efficiency 38, 126–128
    External 127, 128
    Increased loss due to nearby materials 127
    Internal
        Circuit board trace 128
        Wire 128
    Placement 8, 126, 134
    Specialized antenna components 128
Applications xiii, 4, 7, 8, 9, 13–21, 30, 127, 128, 129, 134, 137, 138
Association 11, 35, 59, 60, 61, 71, 74, 76, 88, 98–103, 106, 110, 138, 143
Authentication 8, 90, 92, 93, 143
Automotive sensing 19–20

B

Bandwidth 11, 27, 28, 39, 51, 63, 75, 78, 101, 117, 118, 126, 134, 138, 143, 144, 146
    Allocation 32
    BPSK low rate 27, 40, 41, 44, 48, 137
    O-QPSK 27, 41, 42, 119, 137
Battery life 9, 21, 93, 100, 109, 126
    Cost considerations 25, 146
    Effect of duty cycle on 25, 26, 123, 125
    Effect of standby power consumption on 124
    Effect of time base on 120
    Effect of warm-up time on 118
Beacon 64, 78, 89, 100, 122, 124, 144
    Effect on battery life 93, 100
    Frame 33, 34, 82, 85, 86
    Management 81, 82
    Use in network discovery 100, 106
Beacon frame 33, 34, 82, 85, 86
    Assigned slots 63
    Contention 54, 63, 78

FFD only 34
Superframe 32, 34, 35, 60, 62, 63, 78, 86
Synchronization 35, 60, 79, 81, 89
Uses 34
Bellman-Ford routing algorithm 102, 105
Binary Phase Shift Keying (BPSK) 27, 40, 41, 44, 48, 137
Blocking 51, 120, 135
Bluetooth xiii, 4, 5, 13, 19, 134, 144

C

Cable replacement xiii, 4, 13, 15, 19
Carrier Sense Multiple Access-Collision Avoidance (CSMA-CA) 31, 59, 63, 64, 65, 78, 88, 133, 138
Channel access 29, 31–32, 101
Channel assignment 40, 41
Channel blocking 51
Channel capacity 32
Channel filters 28, 117, 118
Channel scan 100
    Active 73, 74, 89, 95, 99
    Energy 73, 133
    Orphan 74, 79
    Passive 73, 74
Channel selectivity 51
Chip 42–44, 120, 121
Cipher, symmetric 90
Ciphertext 90
Clear channel assessment (CCA) 37, 54, 55, 133, 137, 144
Cluster 95, 113
Cluster head 31, 61, 110
Cluster-tree 30–31, 34, 35, 61, 95, 110–112, 113
Coexistence 11, 131–134, 137
Coherence 27, 28
Collision 32, 35, 59
Colocated transceivers 128, 134
Command type 88
    Association request 61, 74, 88
    Association response 88
    Beacon request 73, 88
    Coordinator realignment 79, 81, 88, 89
    Data request 64, 80, 83, 88
    Disassociation notification 88

GTS allocation 77
GTS request 77, 88, 90
Orphan notification 79, 88
PAN ID conflict notification 88, 89
Communication range 9, 48–50
    Free space maximum 48, 49
    Log-normal shadowing 49
Consumer electronics xiii, 4, 9, 13, 15, 16, 17, 19, 137
Consumer Electronics Association 16
Contention 32, 54, 77, 78, 101, 138
Contention access period (CAP) 32, 54, 63, 64, 75, 77, 78, 88, 90, 125
Contention-free period 63, 75, 77, 78, 90
Coordinator 144
    PAN coordinator 21, 29, 30, 31, 32, 60–66, 73, 75–81, 83, 86, 89, 95–111, 125, 132, 133, 134, 144, 147
Cost considerations, architectural features
    Data rates 29
    DSSS 27, 28
    Duty cycle 25, 26
    Error control 118
    Low battery cost 25
    Low-power transmitter 29
    Modulation 26, 27
    Protocol 122
    Relaxed sensitivity 29
CSMA-CA 31, 59, 63, 64, 65, 78, 88, 133, 138
Cyclic redundancy check (CRC) 34, 85, 144

D

Data frame 34, 35, 65–66, 85, 86, 100, 148
Data link layer 6, 59, 144, 146, 147
Data rates 5, 39, 40
    BPSK low rate 27, 40, 41, 44, 48, 137
    Cost considerations 29, 40
    Frequency bands 39, 40
    High rate 27, 29
    O-QPSK 27, 40, 137
Data transmission 67, 134
Despreading 27
Device synchronization 61, 62, 78, 80, 83, 89
DeviceNet 15

Index

Devices
    Full Function (FFD) 30, 34, 60, 61, 73, 82, 86, 89, 95–101
        Network coordinator 30, 82, 144
        Network devices 30, 146
    Reduced Function (RFD) 30
        Network devices 30, 101, 146
Direct Sequence Spread Spectrum (DSSS) 48, 137
    802.15.4 implementation 39
    Benefits 27, 28, 39, 117
    Binary Phase Shift Keyed (BPSK) 44
    Binary Phase Shift Keying (BPSK) 137
    Coexistence with other services 28
    Cost considerations 28, 39
    Despreading 27
    Multipath interference 117
    O-QPSK 137
    Spreading 27, 28, 44, 50, 117
Disassociate 88
Disassociation 59, 74, 76, 138, 144
Distortion
    Adjacent and alternate channels 51
    Intermodulation 51
Duty cycle 37
    Cost considerations 25, 26
Dynamic channel selection 132–134

E

Ease of installation 8, 9, 11, 17, 19
Eavesdropping 107, 108, 111
Encryption
    See Security, Features, Encryption
Encryption algorithm 138
Energy detection 37, 55, 56, 73, 133
Envelope power
    Average 119
    Peak 119
Error control
    Methods 33
European Telecommunications Standards Institute (ETSI) 10, 38
    Rules EN 300 328 132

F

Federal Communications Commission (FCC) 10, 38
    Rules 47 CFR 15.247 131
Fieldbus 15
Filtering 118, 132
Flooding routine algorithm 102
Frame
    Acknowledgment 33, 34, 35, 59, 68, 70, 85–88
    Beacon 33, 34, 82, 85, 86
    Data 34, 35, 65–66, 85, 86, 100, 148
    Integrity check 63, 91
    Lost message 67
    MAC command 34, 35, 54, 85, 87, 148
    Sequence check 91
    Structure 34–36, 84–87
    Synchronization signal 35
Frequency bands 10, 37, 39, 40, 48, 49, 145
Frequency hopping 28, 117
Frequency reference 120
Freshness, sequential
    See Security, Features, Sequential Freshness
Full Function Devices (FFD) 30, 34, 60, 61, 73, 82, 86, 89, 95–101, 106, 144, 146

G

GRAd routing algorithm 103
Guaranteed time slot (GTS) 29, 31, 32, 59, 63, 64, 75–78, 88, 89, 138

H

Half-duplex 27
Handshake 33
Hash algorithm 93
Hierarchy 102, 110, 111
Home automation 17, 19, 113
Home security 17–18
HomeRF 16
HVAC 17, 113

I

IEEE 64-bit address 60, 71
IEEE Std 802.11 14, 132, 145
IEEE Std 802.11i 93
IEEE Std 802.15.3 xiii, 4, 14, 93
IEEE Std 802.15.4 138
    Cost considerations 25–29
IEEE Std. 802.11 3, 119, 133

Implementation
    Dedicated host processor tradeoffs
        121–122
    Single chip/two-chip tradeoffs 120–
        121
INCOM 15
Inductive field 9
Industrial and commercial control and
    monitoring 14–15, 113, 137
Industrial, Scientific, and Medical (ISM)
    band 39, 145, 146
Infrastructure xiii, 4, 12, 25, 143
Installation cost 3, 7, 9, 14, 19
Integrity
    See Security, Features, Integrity
Interference 3, 28, 32, 33, 51, 74, 117, 131,
    132, 133, 134
Intermodulation 51
Interoperability 3, 12
ISO/OSI 6, 37, 59
    Five-layer model 6
    Seven-layer model 6

L

Latency xiii, 16, 20, 21, 25, 29, 30, 32, 35,
    63, 100, 101, 106, 112, 113, 121
Light communications 8
Limited range 11
Link layer 6, 144, 146, 147
Link Quality Indication (LQI) 134
Logical Link Control sublayer (LLC) 59,
    146
Log-normal shadowing 49
Low cost 5, 25–29, 122, 127, 132, 137
Low power 5, 9, 25–29, 131–134
Low-Rate Wireless PAN (LR-WPAN) 3,
    146
LR-WPAN 5

M

MAC
    Command
        MAC command frame 34, 35, 54,
            85, 87, 148
    Common part sublayer (MCPS) 67, 86
    Data service 70, 71
    Frame structure 34, 84–87
    Header (MHR) 34, 85, 87
    Management service 71
    Payload Data Unit (MPDU) 34, 56
    Service Data Unit (MSDU) 34, 71, 103
    Sublayer Management Entity (MLME)
        67, 71, 95
Maintaining PAN 88
MCPS
    DATA 86, 90
Medium Access Control sublayer 32, 35,
    51, 56, 59–93, 98, 101, 146
Message authentication code (MAC)
    See Security, Message authentication
        code (MAC)
Message integrity code (MIC)
    See Security, Message integrity code
        (MIC)
MLME 67, 71, 73, 95
    ASSOCIATE 74, 75
    BEACON-NOTIFY 82
    DISASSOCIATE 75
    GET 72
    GTS 71, 75, 78
    ORPHAN 71, 79
    Reset 72
    SCAN 73–74, 79, 88, 95, 132
    SET 72
    START 81, 82, 89, 96, 98, 99
    SYNC 79, 80
    SYNC-LOSS 79, 81
    TRX-STATE 72
Mobile phone xiii, 4, 9, 19
Mobility 8, 40
Modulation 10, 37, 40, 41, 42, 44, 117, 119
    Cost considerations 26, 27
Multihop 10, 11, 18, 60, 71, 95, 101, 112,
    113, 147

N

Narrowband 28, 117, 134, 146
Network
    Components 29
    Device 29, 30, 31, 32, 35, 60–66, 75,
        77, 87, 88, 99–112, 125, 138, 146,
        147
    Formation 61, 73, 82, 95, 98–100, 110,
        132, 137

Index

Layer 11, 18, 55, 59, 61, 66, 71, 72, 95, 101, 132, 138, 144, 146
Multihop 10, 11, 18, 60, 71, 95, 101, 112, 147
Peer-to-peer 18, 31, 59, 61, 66, 71, 82, 95, 101–112, 138, 147
Topology 11, 30, 113
    Cluster 95, 104–109, 113
        Routing algorithm 108
    Cluster-tree 30–31, 34, 35, 61, 95, 109–112, 113
        Routing algorithm 111
    Mesh 20, 21, 60, 95, 113
        Routing algorithms 102–105
    Multihop 113
    Peer-to-peer 113
    Star 34, 35, 59, 60–61, 64–66, 71, 95, 113, 125, 138, 148

O

Offset Quadrature Phase Shift Keying (O-QPSK) 27
Original equipment manufacturer (OEM) 11, 121, 122
Orphan device management 79
Orthogonal signaling 27
OSI Physical Layer (PHY) 6, 37–56

P

Packet
    Acknowledgment 56
Packet structure 34–36, 55–57
Packet synchronization 34–36
    Synchronization
        Packets 41
Packet types
    Acknowledgment 35
    Beacon 34–35
    Data 35
    MAC command frame 35
PAN 147
    ID Conflict 88
    ID conflict 81, 89
    Identifier 89, 99, 100, 102
PAN Information Base 53
Parent-child relationship 31, 105, 106
Payload 21, 34, 103, 147

Beacon 86, 91, 106
Data 35, 56, 82, 91
Encryption 91
MAC command 34, 35, 84, 87, 91
PC peripherals 15, 16, 19, 101, 113, 128
PC-enhanced 18
Peak-to-average power 27
Peer-to-peer 18, 31, 59, 61, 66, 71, 82, 95, 101–112, 113, 138, 147
Personal computer 15, 18, 128
Personal Digital Assistant (PDA) 4, 9, 13, 15, 16, 18, 21
Personal healthcare 18, 19
Personal Operating Space (POS) 4, 147
Physical Layer (PHY) 37–56, 147
    Bands
        Frequency ranges 39
        Market considerations 38–39
        Regulatory considerations
            European region 38, 41
            North American region 38, 41
            Worldwide service 38–39, 41
    Bandwidth 39
    Data service 51, 52
    Header 34, 36, 55
    Management service 53
    Packet structure 55–57
    Payload 55
    Protocol Data Unit (PPDU) 55–57
        Header 56
        Payload 56
        Preamble 55
        Start-of-frame 55
        Synchronization header 34, 55, 55–56
    Service Data Unit (PSDU) 34, 56
    Specifications
        Blocking 51
        Transmit power 48
PIB
    See PAN Information Base
Plaintext 90
PLME 51
    CCA 54, 55
    ED 55
    GET 53
    SET 53

Index

Polling 79, 80, 83, 100, 138
Power consumption 3, 5, 72, 112, 117, 119, 120, 122, 124, 126, 135, 137, 138, 148
Power cycling 9
Power spectral density 28
Precision agriculture 20–21
Primitives
    Confirm 52, 54, 70, 71, 78, 81, 83, 95
    Indication 52, 70
    Request 52, 70, 71
    Response 52, 70
Processing gain 28, 117
Propagation 3, 8, 11, 17, 49
    Model 49
Protocol 15
Protocol stack 17, 31, 59, 121
    Architecture 67, 68
Pseudo-random sequence 27, 42
    BPSK 44
    O-QPSK 42

Q

Quality of service 3, 4, 29
Quasi-orthogonal 41
Queuing 32, 71

R

Radio frequency 9, 10, 17, 131
Range 5, 10, 48–50
Realignment 79, 81, 89
Receiver characteristics
    Selectivity 50, 118
Reduced Function Device (RFD) 30, 98, 101, 146, 147
Regulations 3, 7, 10, 18, 26, 37–40, 131, 132
Regulatory agencies 10, 28, 33, 38, 48
Remote keyless entry 17, 19
Replay attacks
    See Security, Replay attacks
Routing table 12, 61, 102, 105, 108, 109, 111

S

Secured mode
    See Security, Modes, Secured mode
Security 3, 90–93

Advanced Encryption Standard (AES) 93, 143
Features
    Access control 59, 90, 92, 93
    Encryption 90, 92, 93, 148
    Integrity 63, 91, 92, 93, 138
    Sequential Freshness 91, 92, 93
Information 8
Length of transmitted frame 93
Message authentication code (MAC) 93
Message integrity code (MIC) 91, 138
Modes
    Access control list (ACL) mode 90, 92
    Secured mode 90, 92–93
    Unsecured mode 90, 91
Replay attacks 91
Suites 92, 93, 147
Security services 90–93
Self-healing 61
Self-organization 12, 60, 61, 103, 143
Sensitivity 10, 48, 49, 126, 134
    Cost considerations 29
Sequence number 87
Service 29
Short-range 3, 8, 9, 11, 13, 25
Signal-to-noise ratio (SNR) 27, 28, 134, 135
Spread Spectrum
    Direct Sequence (DSSS) 37, 39, 44, 48, 50, 117, 137
    Processing gain 28, 117
DSSS
    Orthogonal signaling 27
    Frequency hopping (FHSS) 117
Spread spectrum 10
    Direct sequence (DSSS) 40
Standby power 26, 27, 124
Starting PAN 88
Stick-on sensor 13, 30
Superframe 32, 34, 35, 60, 62, 63, 78, 86, 88, 101, 138
    Boundary marker 35
    Structure 62–64
Symbol to chip mapping 43
Synchronization 9, 21, 41, 61, 62, 67, 72, 78, 80, 81, 83, 89, 100, 117, 144

Chips 55
Control 79–81
Data bits 55
Frames 35, 60, 81
Header 55–56
Packets 34–36, 67
Symbol 36
Synthesizer 118
System architecture, for low cost
    Data rates 29
    DSSS 27, 28
    Duty cycle 25
    Error control 118
    Low battery cost 25
    Low-power transmitter 29
    Protocol 122
    Relaxed sensitivity 29
System-on-a-chip (SoC) 29, 132

T
Telematics 19
Time base reference 120
Tire pressure monitoring system 13, 19
Toys and games 15, 18, 19, 128
Transmission scenarios 67–68
Transmitted noise 134, 135
Transmitter
    Cost considerations 29
    Output power 29
        Frequency bands 48
        Modulation methods 27

Type approvals 37

U
Ubiquitous 25
Ultrasound 9
Unlicensed
    Bands 10, 11, 25, 28, 59, 131
    Cost considerations 25, 28
    Service 28, 37
Unlicensed Bands 39
Unsecured mode
    See Security, Modes, Unsecured mode
Untethered xiii, 9, 13, 25

V
Virtual wire 13, 19

W
Wideband FM 27
Wireless hub 13, 15, 21
Wireless sensor 8, 13, 14–15, 17, 20
Wireless Sensor Network 7–12, 104, 108, 137
Wireless Sensor Networks 113
WLAN 4, 5, 40, 59, 117, 120, 131, 132, 148
WPAN 4, 13, 40, 51, 59, 117, 120, 131, 134, 135, 145, 148

Z
ZigBee Alliance 6, 7, 16, 138, 148